EMPIRE SHOWCASE

EMPIRE SHOWCASE

A History of the
NEW YORK STATE FAIR

Henry W. Schramm

North Country Books, Inc.
18 Irving Place
Utica, New York 13501

Copyright 1985
by Henry W. Schramm

ISBN 978-0-932052-39-1

All Rights Reserved

No part of this book may be
reprinted without written permission
of the copyright owner.

FIRST EDITION

North Country Books, Inc.
18 Irving Place
Utica, New York 13501

About the Author

A former newsman with *The Post-Standard*, Syracuse, and the *Binghamton Press*, Henry W. Schramm calls on his experience as a reporter and State Fair publicist in developing his history of the Fair. He has authored several books, including two histories of the Syracuse area with the late William F. Roseboom, *They Built a City*, and *Syracuse From Salt to Satellite*, and numerous magazine articles. A one-time correspondent for *Business Week*, the author has received his B.A. and M.A. degrees in history and journalism from Syracuse University.

Currently he is Senior Public Relations Officer at Key Bank of Central New York.

A native of New York City, he served in the navy during World War II and as a Signal Corps photographic officer in Germany during the Korean War.

About the Photographer

Albert Y. Edison, a commercial photographer since 1946 following World War II army photographic experience in Europe, has been associated with the State Fair since 1949, and served as Chief of Photo Operations there for a dozen years. He attended Syracuse University before entering the photography field where he specializes in public relations and advertising photography.

A licensed pilot, who mixes pleasure with business as an aerial photographer, he is qualified for both powered aircraft and sailplanes.

Preface

The jumbled acreage of architectural oddities that comprise the New York State Fairgrounds form a strange little nation which blooms briefly each year in robust colors, sounds and smells. Then, just as we begin to get used to it, it's gone again.

Over the decades it's been bent, folded, spindled and mutilated. Yet somehow each year it's there again, putting on a brave new face, a true, though sometimes tenuous link with the essence of Americana. It is politics, regional rivalries, technical and agricultural progress and raw drama. It's also partly "big time" and partly "honky-tonk." Some people have lived for years within sight of the grounds, but never been inside. Others annually choose their vacations around the magic dates.

And surprisingly, no one had ever tried to put its 140-year history down on paper. I became acquainted with the fair when I covered the pathetic little "limited" edition in 1948 as a reporter for *The Post-Standard*. Somehow this show convinced the powers-that-be that a full fair should once again be instituted, and on the same location as the pre-war event. Later I served on the publicity staff, churning out press releases, home town features on prize cabbages and cows, and escorting "celebrities" to and from press conferences and the airport.

I was privileged to have as a teacher Doug Johnson, one of the state's most knowledgeable individuals on the fair. I worked alongside other gentlemen like Roy Fairman, "Mr. State Fair" for *The Herald-Journal* for half a century; Howard Merritt of *The Post-Standard*; photographers such as Herm Borzner, Art Cornelius, Bob Johnston, Frank Monell, and Joe Rensen who died in 1959 while working the fair for *The Herald-Journal*.

I was fortunate as well to have made the acquaintances of such Fair people as Professor Erl Bates of Cornell who was a major factor in development of the Indian Village on the grounds; Ira Vail, the old-time auto racing star who promoted the fair's races for many years, and Directors Bligh A. Dodds and Harold Creal.

More recently Norm Rothschild and current director Tom Young and his public relations associate, Joe LaGuardia, have been most helpful in providing guidance and source materials. Art Gabriel, executive director, and Jack Delaney of the Syracuse Automobile Club, made available that organization's files on early fair races. The collection of Al Edison, who served for many years as official fair photographer, was of primary importance.

Once again Richard Wright and Violet Hosler of the Onondaga Historical Association and their assistants were invaluable in developing this history.

Gerald Parsons of the Onondaga County Public Library's Historical and Genealogical Section patiently sorted through restricted shelves and files to meet my needs.

The Canal Museum's Todd Weseloh again provided a valued source as did the Greater Syracuse Chamber of Commerce through its president, Erwin G. Schultz and Susan Snook.

And the various folks at the Syracuse Newspapers — people such as C. W. McKeen and his photography staff; Sports Editor Arnie Burdick of *The Herald*; Dick Case and Joe Ganley, and librarians Evelyn Clayton and Liz Bahoth were all most cooperative in providing needed information and materials.

Photographic postcards of the fair during the early days of the century were kindly loaned to me by a number of individuals including Edward J. Beiderbecke of Williamson, Mrs. Dorcas Dean of East Homer, William Delaney of Elbridge and Bruce Tracy of Syracuse.

I am indebted to Red Parton, former public relations director for the Syracuse Nationals of the National Basketball Association for guidance in the section devoted to the year-round use of the grounds, and to promoter Glenn Donnelly of the Drivers Independent Race Tracks (D.I.R.T.) for data on the Schaefer events.

Out of town historical associations, among them the Chemung County Society; the Oneida County Historical Association; the Dutchess County Historical Association and the New York City Public Library, provided guidance in searching out materials on the earlier fairs before the locale was permanently sited here in 1890.

The personnel at the SUNY at Morrisville library, including Stella Colleen, and Professor Don Smith of the Agricultural Department were especially helpful in making available records of the State Agricultural Society.

I have tried to assimilate this great wealth of material and present it in a way which is both informative and entertaining — in short, the way the fair tries to present its annual message.

Henry W. Schramm

Table of Contents

PREFACE

CHAPTER I   ON THE ROAD 1

                        The First State Fair
                        Heritage of the Early Fairs
                        On the Move
                        A Return to Syracuse
                        Elmira's Fair
                        On the South Side
                        The War Years
                        Seeking the Permanent Fair

CHAPTER II   THE FAIR COMES HOME 22

                        A Site is Chosen
                        Opening Day
                        The Case of the Rich Rooster
                        Gridiron Action
                        Continued Growth
                        The State Takes Over
                        Turn of the Century
                        The Syracuse Mile
                        McKinley
                        Something Rotten in the Apple Display
                        "The President is Dead"
                        Theodore Roosevelt at the Fair
                        Other Attractions
                        Day of the Automobile
                        The "Dip" Detectors
                        Mystique Krewe
                        Arrival of the Air Age
                        For "A Greater Fair"
                        First Construction Phase
                        Spectacular and Scientific
                        A Boyhood Look

CHAPTER III   THE DAYS OF THE PALE HORSE 64

                        When the Senators Played the Reds
                        The First Air Race
                        Pompeii Destroyed — Again
                        The Mid-Teens
                        The Child Plague
                        The Fair Becomes an Army Camp
                        The 1917 Fair
                        The 1918 Reopening
                        Our Greatest Catastrophe

CHAPTER IV	FLAMBOYANT YEARS	87

        1919
        The Saga of Tex McLaughlin
        The Coliseum
        Another Year of Tragedy
        The Hambletonian
        An Indian Village
        Endurance Flight
        Back to the Mile
        The 1930's
        It Becomes an "Exposition"
        Planning the "Monster Fair"
        The Late 30's
        A Lament for Worthy Souvenirs

CHAPTER V	A TIME TO REBUILD	114

        After Pearl Harbor
        A Time to Rebuild
        A New Home?
        The Interim Fair
        The Full Fair Returns
        A Question of Osmosis
        A Great Deal of Bull

CHAPTER VI	THE MODERN ERA	141

        "EXPO"
        The Making of a Blood Brother
        The Seventies
        Democrats Take Over
        Rocky Times
        A Period of Growth
        Rambling the Grounds
        The Midway Mile
        The Bottom Line

CHAPTER VII	OFF-SEASON	168

        "Bring Back the Hambletonian"
        Fair on Ice
        Hosting the Orange and the Nats
        Dirt Tracking

CHAPTER VIII	IN CONCLUSION	175
BIBLIOGRAPHY		177
NOTES		179

CHAPTER I

On The Road

The striped tents and bright pennants blended excitingly into the lush English meadowlands. Squires and working hands alike enjoyed time off from pasture and barn as they traded stories at plowing matches and horse races. Even better was the enjoyment of a pint of bitters and a generous portion from the delicacies available at the refreshment tents. One could even learn how to improve his farming techniques by listening to the "agronomists" lecturing in the assembly pavilion.

The year was 1777. The occasion, the great Bath Agricultural Fair, a uniquely English phenomena.

Several thousand miles to the west, their countrymen were trying to decipher the crude charts of the Colony of New York's outer wilderness.

The dense forests, swamps and lack of roads offered little incentive for soldiers trained in the Continental traditions. Even the few names dotting the maps were foreign and unnerving: Ticonderoga, Saratoga, Oswego. . . truly a Siberia for gentlemen who had to make do with natives and Hessians as comrades in arms.

The Red Coats didn't stay too long.

But the concept of the English agricultural fair, imported at about the same time, remains a healthy heritage from the Colonial period despite, or perhaps because of, our space age electronic civilization.

Although a Trenton District Fair was held in the Colony of New Jersey as early as 1745 under the patronage of King George II of England, it never became a state fair until long after New York had inaugurated its own yearly exhibition.

It's altogether fitting that the first to adapt the fair as a statewide event would be New York, with its close ancestral ties to England and its early dependency on farming for its very existence. It's even more fitting that the first American State Fair be held in the center of the state.

What is amazing is that it took less than a man's lifespan to transform a wilderness into a civilized community suited to holding a gathering of tens of thousands of visitors, arriving by modes of transportation not even dreamed of by the men of '77.

At the turn of the 19th Century, the outer limits of the former Colony, once one left Albany, were quite primitive. Movement west of the capital usually meant a rustic journey along Indian trails or the Mohawk River. Travel by horse and wagon wasn't possible until the post-Revolution years when veterans received tracts of upstate land as a thank-you for wartime services, and then built turnpikes to get to them.

In 1792, early settlers in Onondaga County, for example, carried the components for the region's first sawmill over their shoulders.

And it was next to impossible to move the precious barrels of salt, the mainstay industry of Central New York, from the processing points near Onondaga Lake to eastern markets.

Life was difficult. Dozens of early settlers died each year from the fevers borne in the mosquito-infested swamps. Cold winters and lack of anything but the barest essentials left most residents in a perpetual state of poverty.[1]

Consequently, most Upstate New Yorkers were a bit behind their New England cousins in developing a taste for shows and exhibitions. But even in those times farsighted men were laying the groundwork for the future state fair.

In 1791 New York's Society for the Promotion of Agriculture, Arts and Manufacturers was organized, and two years later received its legislative charter. One of the Society's first moves was to publish papers on agriculture, a very real service to the information-starved farmer.

Founders of the Society were themselves farmers who had the traditional problems of the countryman: fertilization, soil exhaustion, drainage, insect depredations, improving the breed and drought. They saw the need to share their experiences and to bring in experts to come up with answers.

Their perception preceded by several years George Washington's statement to Congress in 1796 that a National Board of Agriculture be founded.[2]

Next on stage was Elkanah Watson, a city dweller from Albany who bought a 250-acre farm near Pittsfield, Mass., just 36 miles to the east. As a gentleman farmer, he acquired a pair of Merino Sheep, the first-ever to appear in the Berkshire region. He decided that the unusual beasts deserved center stage.[3]

So many people stopped by his little show under a tree in the public square that Watson decided to initiate a full-scale event. By 1810 the Berkshire Agricultural Society was sponsoring its cattle show in Pittsfield annually.[4]

Many familiar fair attractions, among them prize contests for the best jellies, pickles and mincemeat vied with livestock competitions to make the event a truly family affair. Later quilting bees and horse races became standard attractions.

Several years later Watson sold out his acreage and returned to Albany. But the enthusiasm generated by the successful showings impelled him to push for the formation of a State Board of Agriculture and attendant government support for a series of county fairs.

Watson soon became the "Johnny Appleseed" for fair activities in New York State (even though Bath, N.Y. is said to have had a fair and horse races as early as 1793). In 1816 he organized and promoted the Otsego County Fair in Cooperstown with premiums totaling $100.

On February 4, 1818, the Agricultural Association of Cayuga County was formed, and starting the following October Auburn was the scene of many county fairs.

Onondaga County's first fair was on November 2, 1819 under the auspices of the newly-founded County Society at Onondaga Hill. Premiums were just $200.

The early fairs were developed along the Berkshire plan, which required at least two days of activities; establishment of an executive committee to manage the affair, and assurance that neat, permanent pens and proper places for exhibiting animals and goods be available. The rules stressed that premiums for the following year's fair should be announced not later than November so that entrants would have almost a full year for preparation.

Recommended were pastoral balls, an innocent type of festivity which would encourage full community interest for townsdwellers as well as for farmers, and to enlist the support of the womenfolk.

Proper fanfare was a must:

"Let the morning of each day be ushered in with the ringing of bells, firing of cannon and the display of farmers' flags to rouse the dormant energies of the community."[5]

It wasn't until April 26, 1832 that the New York State Agricultural Society, the direct parent of the State Fair, was finally established in Albany by an act of incorporation by the legislature, largely through the efforts of Elkanah Watson and another strong advocate, James LeRay DeChaumont of Jefferson County, who became the first president.[6]

Among activities considered within the scope of Society encouragement were stock husbandry, tillage husbandry, horticulture and a catch-all category, "the household arts." This last group included such handicrafts as the making of household fabrics; the rearing of silk worms (a fairly short-lived experiment in Upstate New York, although actively pursued for a time in the Auburn area) and the preparation of "domestic wines."[7]

THE FIRST STATE FAIR

Even though the public didn't seem overwhelmingly supportive of the idea of a state fair, the legislature came through with $8,000 in 1841 for "promotion of agriculture and household manufactures in the State" through an annual fair. The village of Syracuse, considered the center of farming interests in New York, was selected for the two-day event.[8]

This was formalized when the Executive Committee acted in April, with Henry S. Randall of Cortland motioning a resolution "that the New-York state Agricultural Society will hold its annual fair in the village of Syracuse, on Wednesday and Thursday, the 29th and 30th days of September next." It was seconded by Alexander Walsh of Rensselaer.

At the same meeting a sub-committee was appointed to develop a premium list.[9]

The village's selection was logical. Syracuse was a centralized port on the Erie Canal and a way-station on the developing series of railroad lines tieing Albany to Buffalo. A network of toll roads and turnpikes provided efficient stage line connections with cities and towns in all directions.

In 1841 Syracuse's population was 11,000, many employed in the production of salt, with vast fields of evaporation sheds and salt blocks extending along the shores of nearby Onondaga Lake. "Downtown," once a swamp, had been drained. Churches, retail stores, offices and several hotels such as the four-story Syracuse House, provided an aura of bustle.

The town could not be construed as "stodgy" by any stretch of imagination. Taverns were commonplace and the residents enjoyed their gambling, including a love for racing horses.

The sport flourished in Onondaga County from the time a competitive course could be etched out of the forest. In the late 1820's a mile track was developed in the western part of the village, with a race for the then almost unbelievable sum of $1,000 attracting horsemen from all over the state. They saw Paul Pry, a Syracuse horse, narrowly defeat Salt Point John from the neighboring village of Salina.

For the pure sport of it, an amateur driving club of "prominent business and professional men" conducted a series of races. It's probable a few bookies watched from the sidelines.

Gambling developed as a bit of a problem and there was great concern about the establishment of a horse track to coincide with the fair, and the lawless element it would attract. (In 1841 fully ten per cent of the population were considered by some community leaders as part of the underworld).

On August 26, 1841 the village's trustees called a mass meeting "for two days' hence" to take action against the holding of races in the village. Not only were three extra constables hired for the occasion, but a vigilance committee of 50 was established. Two hundred townspeople volunteered![10]

As to the fair itself, a suitable site was found midway between Syracuse and Salina, close to transportation and within walking distance for residents and "downtown" Syracusans.

The old Onondaga County Courthouse occupied a section of the grounds, with the jail nearby in the southwest corner of a large, unsettled space bounded today by North Salina, Division, Townsend and Ash Streets. A pleasant grove of pine trees stood to the northeast of the area.

This grove became the center for animal displays while the two-story brick courthouse with its statue of justice atop the dome, was chosen for the exhibition of all sorts of farm produce, implements and manufactured goods.

Even though a tremendous fire and explosion several blocks from the grounds a month before the scheduled opening resulted in death to 30 persons and injuries to hundreds of others, the people of Syracuse gamely carried on. The fair opened on time.

With Mayor Harvey Baldwin, Philo Rust and M.D. Burnett, a leading Syracuse citizen, among those in charge of arrangements, the two-day event turned out to be a rousing success . . . if you like overwhelming crowds jamming overtaxed facilities. One large room on the courthouse's second floor contained many of the main attractions, and from morning to night a constant stream of viewers undertook the perilous journey up the stairs past the descending multitudes.

The Society, in its annual report stated that these "were proud days for the State of New-York, commencing, as they did, a new era in our agricultural career; giving a well founded hope of many succeeding and still more triumphant gatherings of the bone and muscle of our country."[11]

Some indication of the success can be imagined in that attendance was large, with estimates ranging from between 10,000 and 15,000 persons, principally farmers, representing every county, plus many from adjacent states. And this, even though the weather was bad.

The weather, it was said, largely "prevented the attendance of ladies, although enough were present to show that in everything that regards the public welfare, man may be sure of the cordial support and approbation of woman."[12]

The show also attracted a tremendous number of exhibitors. On the day before the fair opened, a special train of 25 cars, crammed with choice livestock from Albany and Hudson River communities, plus a passenger wagon, journeyed from Albany to Syracuse.

There would have been even more cattle if a canal boat carrying the prize herds of Ezra Cornell (later to found Cornell University) hadn't run aground.[13]

Considering the modes of travel during the period, the overall turnout was remarkable.

Although the train was coming into its own by the time of the first fair, with a chain of disconnected lines extending from Albany, many early fairgoers still depended on the stagecoach over the shorter hauls and from the north and south, facing up to the bone-rattling journeys over scotch macadam, plank and log roads. The speed of the swaying vehicles was nothing to brag about, averaging three to four miles an hour when the frequent wayside stops for meals and refreshments were factored in.

Not that rail travel was overly pleasant. Aside from the constant danger of derailment or being set afire from embers from the engine stack, passengers from Albany had to change cars at Schenectady and at Utica before arriving in Syracuse. Those coming from the west had to switch at Batavia, Rochester and Auburn.[14]

Several exhibits at that first Syracuse fair attracted special attention. Included were three yoke of oxen from Ontario and Onondaga Counties, weighing 18,000 pounds. Philo N. Rust, proprietor of the Syracuse House, the area's leading hotel, owned a pair. He slaughtered one of the oxen which weighed 2,750 pounds and presented a dressed quarter to the Agricultural Society. The quarters alone weighed 1,784 pounds!

Speeches were important to early fairs. The Presbyterian Church was filled in the evening to listen to President Eliphalet Nott, D.D., of Union College speak on "The Dignity of Labor."

In his speech, Joel Nott of the Society paid special tribute to DeWitt Clinton who's hand had "first put the ball in motion" leading to the Agricultural Society's establishment, and "he not only increased its produce, but opened a highway (the canal) to market."[15]

Mayor Baldwin responded to the congratulations of the visiting speakers by eulogizing the press as the "matchless engine which stirs and lifts up the whole body politic as one voice that speaks with a thousand tongues to a whole nation at once."

Almost every tavern and inn held special dinners to accomodate the tremendous crowds. The draughty wooden railroad station in what is now called Vanderbilt Square on Washington Street, was the hall for reading reports of premium awards, and was crowded for two days with anxious competitors and the curious.

Topping off everything in the thinking of most farmers was the plowing contest. This import from the early English fairs blended strength, skill and equipment in a competition which townspeople and countrymen alike could understand.

The contest was held on the J.H. Johnson farm near today's West Onondaga Street, then referred to as "the great plain of the Onondaga Valley," with oxen providing the pulling power. The ground was hard and dry, "falling to pieces when turned up by the ploughs, rendering it difficult to make a clean furrow, or show the precise manner of the working of the implement."[16]

A cold drizzle was falling, but the crowds rimmed the entire perimeter.

Despite the difficulties, Howard Delano of Mottsville came away with plowing's first prize.

President Nott reviewed the accomplishments of this initial event at the Society's meeting on January 19, 1842. He declared the Syracuse Fair "though a most unusual, was nevertheless a most successful experiment;" that it exceeded their most sanguine expectations.[17]

HERITAGE OF THE EARLY FAIRS

A visitor said, "let it be noted that the show of agricultural equipment was most satisfactory; that the numbers of threshing machines, horse power, straw cutters, fanning mills, plows, harrows, cultivators, etc., was very great and exhibited much mechanical ingenuity and skill."

One official stated later:

"Our best plows are now propelled through the greensward by a draft of 300 to 500 pounds and two or three cattle are competent to turn the most stubborn soils, whereas nearly double the power was required formerly," a situation directly related to the lessons taught by the fair's exhibitors.[18]

For the next half-century the event was like a traveling circus.

The second site was the state capital in Albany, home of the Society and a logical place to showcase to the governor and others likely to encourage greater appropriations for an activity which had proven itself in its initial performance in the hinterlands.

With more than 600 domestic animals on display and a tremendous showing of agricultural equipment including more than 40 plows, the fair was considered a success. It, too, finished in the black with receipts of $345.74 after expenses.

On the final day, then-Governor William H. Seward appeared as guest orator. He called for improving professional education in farming, decrying the fact many considered an inferior education sufficient for those destined to work the soil.

The following year the venue was established in Rochester, where for the first time the 10 acres of grounds was enclosed by a high wooden fence and an admission charge (12½ cents) was made "as a reasonable mode of defraying the expenses of the Society." Some 20,000 visitors crowded trains, stages and canal boats to get there.[19]

Daniel Webster was expected, but was unable to appear so former Governor Seward substituted as keynoter. Other guests included former President Martin Van Buren.

In his talk, as read by Seward, Webster prophetically described the importance of New York as a leading agricultural state. He told the hometown audience that "New York City has been brought very near your doors. The great emporium of this continent lies close before you. You are rich in your home market — a market of purchase and sale. All New York is at your feet. You can deal with her as if you lived in one of her wards. . ."

The 1844 fair, held September 18 and 19, shifted downstate to Poughkeepsie in the Hudson Valley, with exhibition buildings providing protection from the weather specially erected for the occasion. It was the year when the Society decided to establish an agricultural museum.

Central New York once again hosted the fair the following year when Utica was selected. Led by B.P. Johnson of Oneida, president of the State Society, this Mohawk Valley rail, highway and canal center demonstrated tremendous enthusiasm.

Four exhibit halls, each measuring 100 by 50 feet, were built to hold the principal displays highlighted by a demonstration of the magnetic telegraph. Included in the exhibit pens were 683 animals.

It was estimated a record 40,000 persons attended.

The Utica act was a hard one to follow but Auburn, the home of John Jethrow Wood, inventor of the iron plow, and with native son John M. Sherwood serving as Society president, is said to have held a fair with "excellent success." The mid-September event was held on "Capital Hill," so-called because at the time Auburn had designs on becoming the state's capital.

Saratoga was the focus for 1847, with the fair held on a 16-acre site. Among the visitors were John Tyler and Martin Van Buren, the latter obviously an accomplished fair-trotter by this time.

In 1848 the fair could be said to have come of age, as the Buffalo exposition was acclaimed by adherents as being the greatest fair ever held on the North American continent up to that point. A record 1,452 head of livestock and nearly 5,000 implements were shown.[20, 21]

A RETURN TO SYRACUSE

Once again is was Syracuse's turn.

A site was selected on the top of the James Street Hill on the city's northeast side, above what was known at the time as the White Grounds. It was bordered by what is today Highland Street.

The planning committee went to work to provide an enticing variety of speakers and events. President Zachary Taylor was contacted and agreed to deliver the keynote address. But, he took sick while at Niagara Falls and his place on the rostrum was filled by Vice President Millard Fillmore and the crowd's favorite, Henry Clay of Kentucky, who showed especial interest in the gathering and the community by stopping at the grounds twice during his stay.

A speech of a different sort was given by Professor James F.W. Johnson, F.R.S., of Great Britain who delivered a talk on scientific farming and then conducted a series of technical meetings.

As a change of pace, two prize fighters, "Tom" Hyer and a pugilist named Sullivan, presented a series of exhibition bouts.

A number of pavilions were constructed, including a Manufacturers Hall, a Dairy and a Mechanics Hall. The greatest acclaim, however, went to the Floral Hall, where on September 13 the Floral Ball, the social highlight of the season, was attended by Syracuse's leading families. Flanked by an imposing array of local and imported blooms, gowned women and formally attired men competed in showering attention on Henry Clay.[22]

Although Mr. Clay's visits may have been the high point of an eventful week, the physical high was achieved by a ride in a 50-foot-tall contraption which dominated the James Street entrance to the grounds. It served as a beacon to the 65,000 visitors who traveled the mile and a half up the dry, dusty street by foot, by omnibus or by carriage from the railroad station.

The structure was a great iron and oaken wheel with wooden bucket cars, large enough to carry either four adults or six children aloft from the end of each of the four arms.

The revolving wheel with its cars was carefully counterbalanced, carrying the passengers "comfortably" and safely around the circle of the wheel, enabling the riders to obtain a marvelous view of the newly-chartered city and its suburbs.

This wheel was operated by handpower and a system of ropes. The balance was such that the wheel could be turned by the strength of a child.

Thus the great Ferris Wheel which was credited as being an original invention for the 1893 Columbian Exposition was actually 50 years late, although its huge size, the number of passengers it could carry and its steampower made it a far more spectacular vehicle.

The state fair operator was Samuel Hurst, a Syracusan who developed the wheel idea in concert with James Mulholland, a Scotsman and carpenter who had seen a similar wheel in Edinburgh.

Once set up at the fair, Hurst contracted with "Billy" Humes, an elderly and experienced showman who formerly managed the Syracuse Museum. The ride was an instant success, with the cars often filled to capacity at a fare of a Spanish Shilling (the common currency at the time). After the fair Hurst sold the wheel to Harvey Bennett, a Brewerton hotel keeper, who set it up on Baldwin Island.

In 1850 the wheel followed the fair to Albany and in 1851 a delegation from Rochester had Hurst build a model for them so they could construct one for the fair there.[23]

Administrators of the Albany fair tried again to attract President Taylor. Once more he agreed to attend. But his death intervened and Amos Dean was the orator. Other distinguished guests included the grandson of Marquis Lafayette, delegates from 22 states and representatives from several British-American territories.

The repeat sequence continued in Rochester in 1851 with throngs in attendance to hear the speaking genius of Stephen A. Douglas, Senator from Illinois and the era's master of the debate. Some 100,000 visitors were present for the full fair, inspecting 2,014 head of livestock and 1,000 other exhibitions.

The 1852 fair returned to Utica where the premium kitty by now had grown to $7,500.

Bad weather spoiled the Saratoga encore the following year, keeping attendance down to a point where receipts barely met premium expenses.

The Society recognized by 1854 that even though the migratory system did encourage improved farm operations in areas of the state where it was held, the growth in both attendance and exhibitions required increased expenditures for erecting accomodations.

Thought was being given to locating the fair "at two convenient and accessible points, eastern and western, the exhibition to be held alternatively in each."[24]

New York City became, for a few brief days, the agricultural capital of the state in 1854, the city subscribing $8,000 for the venture.

But even with superb management, well thought out working plans which were lithographed and followed as a blueprint by the organizers, and special care afforded the press, the location at Hamilton Square was not convenient to New Yorkers. For the most part, they went about their daily business with little thought to the agricultural show in the area "where the field ceased, and the street and square commenced." Then too, rain fell for several days.

Farmers also found the Big City a little far for convenience, so their contributions were limited.

As a result, the fair could not be termed successful and was never again held there. Receipts of $9,538 were scarcely more than the average for other fair locations during that decade.[25]

THE ELMIRA FAIR

The Elmira area's introduction to fairs came in September of 1841 when local citizens met to form an agricultural society in Chemung County, with the express purpose of establishing a local fair. But that wasn't the only business conducted. The gathering declared a "war of extermination against our common enemy, the Canadian thistle."

The fair was held the following year, with Mrs. Hannah Wynkoop, then 65, receiving recognition for the best manufactured silk, a pair of stockings, the silk being both raised and manufactured by her. Mrs. Wynkoop also received two dollars for the best bed quilt, "a beautiful article of domestic manufacture."

Meanwhile, Hoffman & Gardner turned out the finest (and most beautiful) fur hat.

In the swine category, the committee described the competition as follows: "These Hogs were truly majestic, and in point of size, form and condition, were so nearly equal that, the committee with difficulty (could either the preference, not being in the list,) [sic] no premiums were awarded, but much credit is due the gentlemen who raised and exhibited these noble Hogs."[27]

A dozen years later Elmira was ready to go for the state fair and in 1855 the first of nine events was held in the city on grounds on the north side of East Water Street midway between Sullivan Street and Newton Creek (the site was later occupied by the Kennedy Valve Plant).

Two special activities during the fair included a formal address in the Speaker's Tent, with Indiana Governor Joseph Wright the orator, and a carbon copy of the successful Syracuse Floral Ball in Elmira's Floral Hall, which dominated the center of the grounds.

The fair was an outstanding success, and opened the way for later such events there.[28]

Watertown held its first state fair in 1856, with a high point the presentation of premiums offered by Horace Greeley ("Go west, young man, go west."), editor of the New York Tribune, for the best farming efforts of young men under 18 years of age.[29]

Buffalo hosted the next year's fair. Even though there was a severe economic depression at the time, the show was well attended, highlighted on October 8 by what was described as "the largest crowd ever convened in the state on such an occasion." As to be expected, among the attendees was former President Millard Fillmore.[30]

Introduced at the Buffalo extravaganza was the concept of steam engines for farming purposes, a radical departure from dependency on horse or oxen-drawn equipment. It was to lead to a major revolution in farming techniques and production.

ON THE SOUTH SIDE

Syracuse's turn came again in October of 1858, with the city becoming the first community in the state to host the fair three times.

A new location was chosen on the city's south side, a mile from the center of town. The 31-acre site was bordered to the west by Onondaga Creek. A race track with a special building to house city and county exhibits were among the features.

According to one account, Syracusans prepared for the show "in a competitive and worrisome frenzy."

City officials made sure the downtown area was spic and span. "Our efficient street commissioner is determined that the sidewalks shall be cleared, in anticipation of the state fair. He has already obtained warrants for some of our merchants. . ." commented a local observer.

The hubbub of the fair caused dislocation of other activities. The newly-formed Syracuse Baseball Club, organized but a few weeks earlier, found it difficult to get enough members out for games because of the fair and the press of business accompanying it.

Many private homes joined the area's hotels and taverns in opening their doors for visitors and, hopefully a quick buck. Mrs. Honor at 58 West Water Street pointed out she could "accomodate 50 persons during the fair with board and lodging at $1 per day or 25 cents per meal."

The railroads, by then America's leading means of transportation, channelled in trains from every direction, carrying passengers and articles for sale or display at the fair. On just one day 71 freight cars arrived with fair consignments.

Daily attendance was close to 20,000, with the crowds tightly bunched on the muddy grounds, making sightseeing difficult. Outside the gate an extemporaneous city popped up, with pocket-emptying tents and stands offering refreshments and amusements of all sorts. Blackface minstrels, tightrope performances, flying horses, swings, fat hogs and museums were everywhere. Exhibitors outdid themselves. Mr. Plowright was a special favorite, a gigantic figure of a man in Highland costume, made entirely of vegetables. The large hall of the Onondaga Society exhibited articles produced locally from a steam engine to a shoe peg.

A showing of Reed organs and melodeons from the M.O. Nichols establishment "filled the whole Hall with harmony," a witness said. "These elegant musical instruments," he continued, "are fast taking the place of the more clumsy and inconvenient thumping music boxes that have so long occupied a large space in the parlors of our music-loving citizens."

Two former presidents, Martin Van Buren and Fillmore, and Governor John A. King slogged through the mud to the speaker's platform.

The most dramatic event was reserved however, for one of the sideshow tents where an Edward Spaulding (known professionally as Mr. Tunere) of Baldwin, N.Y. appeared with a large case and a dog team. The case contained his collection of live rattlesnakes! Mr. Spaulding was a pathetic little figure, a cripple who had to walk on his knees. Surrounded on stage by his props, the specialist began to open his case, but experienced difficulty. A newspaperman viewing the scene called the whole thing "disgusting." Finally, with the case open, Spaulding reached in. He let out a scream. A large rattler had fastened itself onto the palm of his hand. Almost at once the arm began to swell. Spaulding administered some potion of his own to the wound, then scrambled along on his knees trying to find a doctor. A large crowd gathered around him, following along as he bumped across the field. Kinder people intervened and he was assisted to the office of Dr. Hiram Hoyt, a well-known surgeon who's office served as both surgery and hospital. Scant hope was given the little man who's arm by now had become enormous. But the doctors cauterized the wound and gave the patient heroic doses of ammonia. The crowd outside the surgery dispersed, anticipating the worst. But the man was obviously a survivor. By the next evening his symptoms were subsiding and he was eating "like a horse," and ready to resume his "dangerous and horrible vocation."[31]

Although this was the last time Syracuse was to host the fair for more than 30 years, the first small thoughts of a permanent fair in Syracuse were entertained in an editorial in the Syracuse *Daily Journal*. The writer said, ". . . the officers of the New York State Agricultural Society cannot do better than to select Syracuse as a permanent location for its annual fair."[32]

THE WAR YEARS

Even the Civil War failed to eliminate the idea of an annual fair, and Rochester in 1862 and 1864 and Utica, in 1863 and 1865 hosted the event, with reports indicating attendance was "average."

Utica pressed to attract the fair permanently. In 1863 its horsedrawn streetcar line running up Genesee Street from the railway station was rushed to completion especially for the show.

The community even constructed several buildings for annual use. They were described in the report of the State Society as "an advance toward the style of edifice in which agriculture will yet display its annual triumphs."

SEEKING THE PERMANENT FAIR

This, according to the volume, *The New York State Fair — It's Genesis and History,* was the focus that led to the fair being affixed in one location.

Elmira, which had held successful fairs in other years, including an evetn with a one-day attendance of 25,000 in 1869, entered the competition when the State Society came up with a plan in 1872 to locate the fair at specified intervals on grounds with suitable permanent buildings and services.

The Chemung County Board of Supervisors provided $50,000 for a 50-acre site and permanent buildings when the Society agreed to contract with the county for a state fair every three years for a 12-year period, and as often thereafter as at any other location. The Southern Tier community thus alternated with Albany, Rochester and Utica as the fair's home for the next few decades.[33]

Despite the enthusiasm of Elmirans, its location off the direct line from Albany to Buffalo led to its dismissal as a competitor for the permanent site.

Pressures were building in Albany and in Syracuse for establishment of a grounds along the great rail lines.

Elkanah Watson, originator of the American Fair System.
Onondaga Public Library

The first Berkshire "Fair," where Elkanah Watson exhibited his sheep. It was the forerunner of the State Fair concept.
Onondaga Public Library

Harvey Baldwin, mayor of Syracuse, greeted the 1841 Fair attendees. His wife was a blue ribbon winner.
Onondaga Historical Association

Site of the first State Fair at Salina and Ash Streets, Syracuse. Note the statue of Justice atop the courthouse.
Onondaga Historical Association

Syracuse House, center of feasting activities for 1841 Fair, was located near the Erie Canal at Salina and East Genesee Streets.

Syracuse's railroad station in 1841 was locale for announcing exhibit winners at first Fair in 1841. Site is now Vanderbilt Square.

Travel by packet boat was one way of reaching early State Fairs, most of which were along canal routes.
Canal Museum

Show horse, circa 1844.
Onondaga Public Library

The 1846 Auburn Fairgrounds activities are shown in a primitive sketch.

The newly-chartered City of Syracuse as it was when the 1849 Fair was held on James Street Hill.

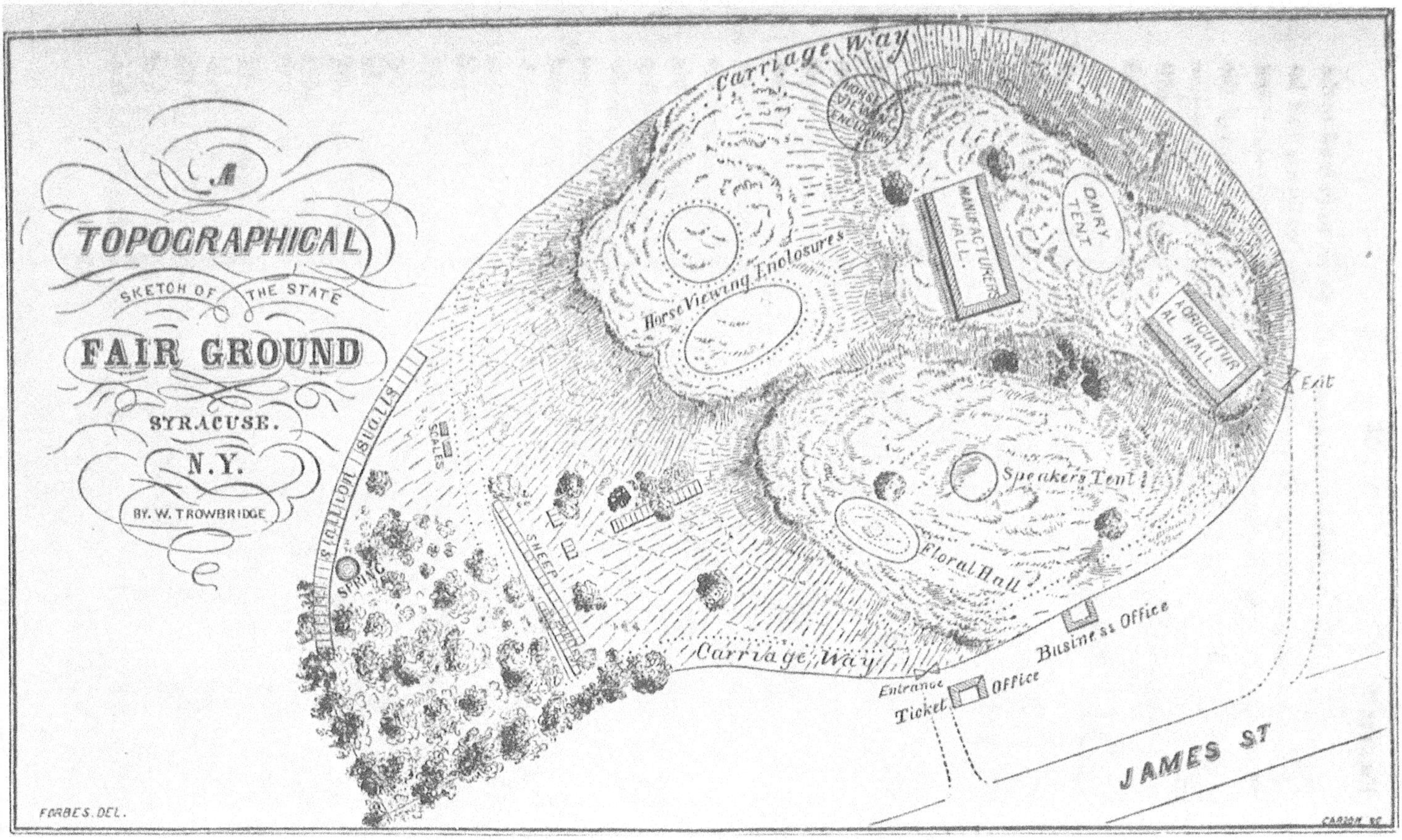

An "aerial view" of the 1849 Fair in Syracuse.

Onondaga Historical Association

The crowds flock up the James Street Hill to visit the 1849 Fair, the second held in Syracuse. Note the "Ferris Wheel" to the left.

Onondaga Historical Association

The Floral Hall, center of attraction, at the Syracuse Fair of 1849.
Onondaga Historical Association

The "Ferris Wheel" which graced the 1849 State Fair atop James Street Hill. It may have been the first such contraption in the New World.
Onondaga Historical Association

1850 Albany Fair.

New York State Fair

A romantic view of an early fair.

New York State Fair

Poughkeepsie Fair of 1844.

New York State Fair

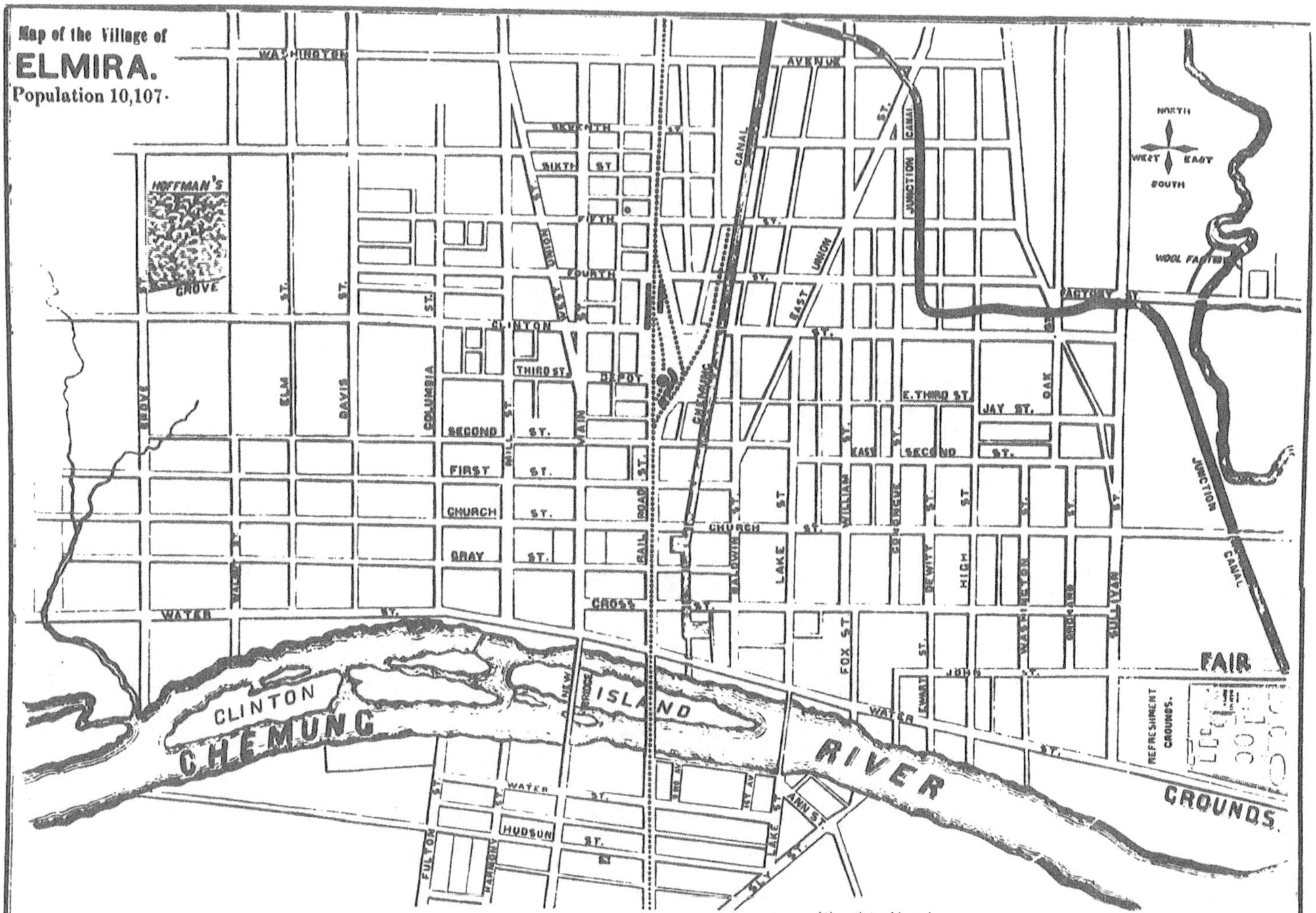

Tucked away in lower right hand corner of Elmira, near the Chemung River, was 1855 State Fair.

Chemung County Historical Association

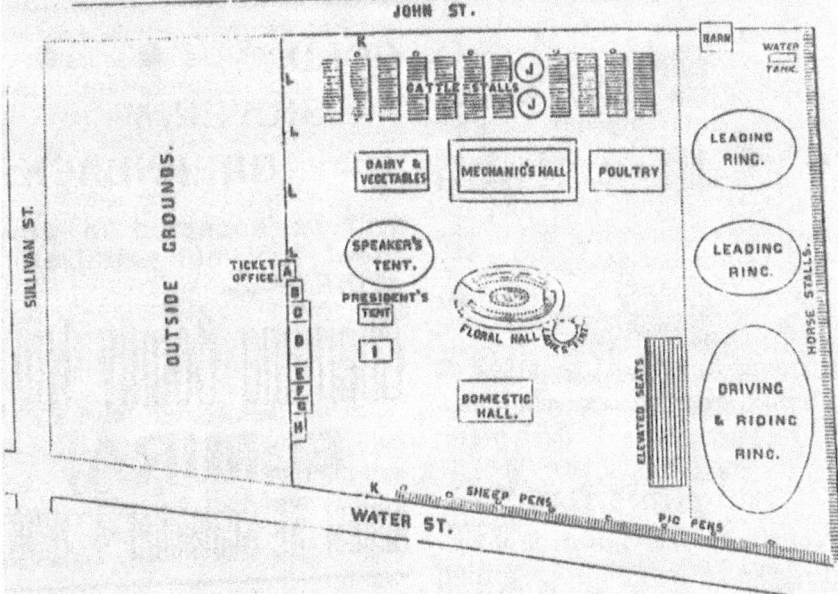

This is the layout for the 1855 State Fair in Elmira. Note the use of a "Floral Hall" as the centerpiece.
Chemung County Historical Association

Mr. Five-by-Seven (round) and other side show performers were heavily advertised attractions at State and County Fairs. Doctors and clergy admitted free.
Chemung County Historical Association

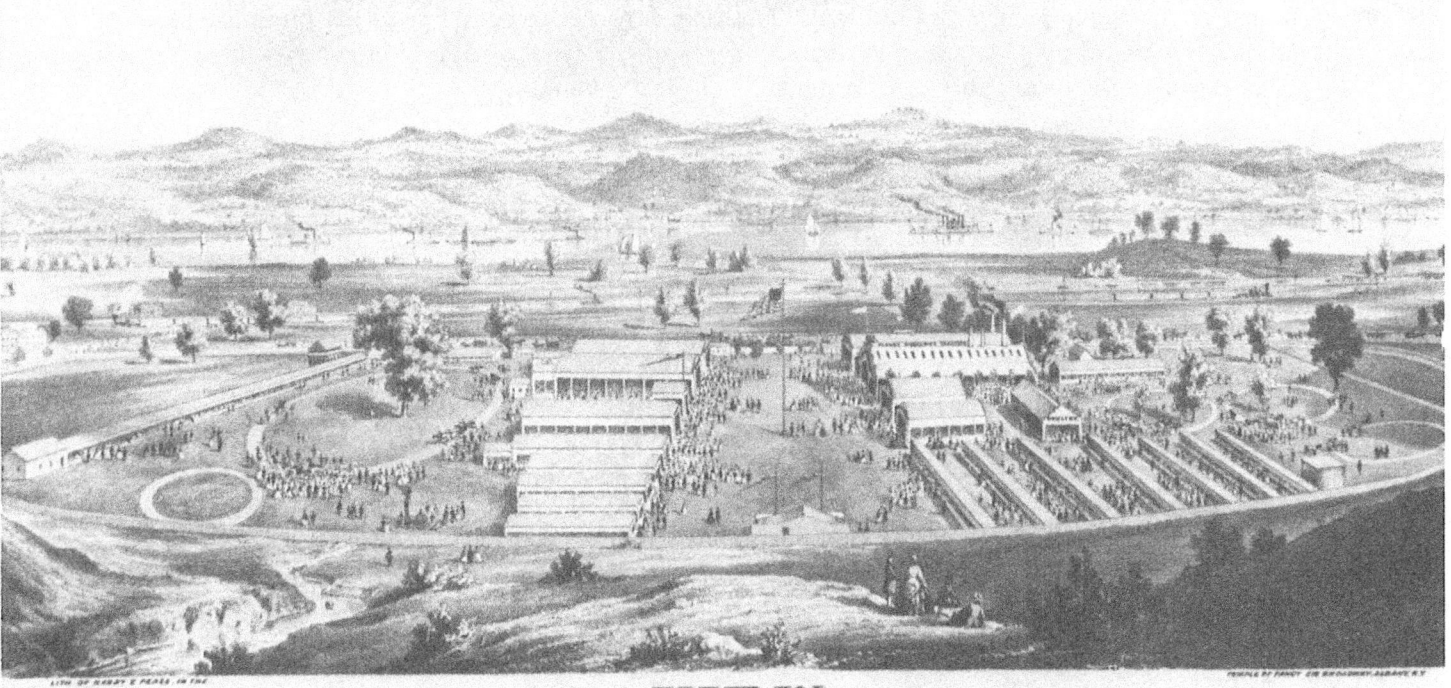

1859 Albany Fair.
New York State Fair

CHAPTER II

The Fair Comes Home

The Syracuse community had grown tremendously since hosting the first state fair. The population reached 88,143 by 1890. It was proud of its new university, founded just 20 years earlier. And, although the salt industry upon which the region had thrived was dwindling in importance, manufacturing, retailing, banking and transportation provided the area with a strong economic base.

Syracuse's leadership believed it had a great deal to offer as a permanent home for the fair.

James Geddes, descendent of one of the early developers of the Erie Canal, downtown Syracuse and the salt industry, must be given major credit for bringing the fair to Syracuse. Elected president of the State Society in 1887, he died that May, but not before he succeeded in setting the stage for locating the fair in one place, Syracuse. Primary reasons were geography and railroad facilities, but there were more.

He commenced the campaign by sending personal letters to each of the 122 life members of the Society, tactfully asking everyone his opinion as to locating the fair permanently in Syracuse. A great majority of the Society members supported the measure. Now it was up to local residents to put together the package. On November 28, 1888 the Syracuse Land Company was chartered to buy and sell land. The real purpose, of course, was the purchase of a fair site. A board of 13 trustees comprising some of the city's strongest citizens — lawyers, bankers, manufacturers and "entrapreneurs" [sic] was named.

A SITE IS CHOSEN

The decision was reached to purchase for $30,000 a 100-acre tract of pastureland comprising the Smith and Powell stock farm and nursery on the city's western suburb, adjacent to both the tracks of the New York Central and the Erie Canal, yet within easy access to the city's business district.

The trustees contacted Syracuse residents and business firms. Early subscriptions came easily. Then things slowed down. The last $5,000 looked impossible. To all that is, but a certain Austin Chase.

Judson Smith, the company's chairman, frankly stated later that without the perseverance of Chase the project would probably have foundered.

Some of the contributors are familiar names to Syracusans even today: Horace White, Dey Bros. Co., Lyman C. Smith, the typewriter manufacturer, John Crouse & Co., D. McCarthy & sons, the Solvay Process Co., Haberle Brewing Co., J.P. Burnet, Attorney Frank Hiscock, E.M. Klock, Jacob Amos, Merrill & Soule, Everson & Co., Francis Hendricks and the James Geddes Estate.[1,2]

The land was bought and offered to the State Agricultural Society at its meeting in

February, 1889. The only condition was that the Society should begin holding fairs on the tract within three years, and should hold a fair there annually thereafter.

The Society's executive committee accepted the offer and went a step further, buying an additional 13½ acres.³

With a permanent site in its possession, the Society went ahead. Livestock buildings were constructed and a half-mile race track laid out.

Just one piece of unfinished business remained. The Society lived up to its earlier commitment and the last of the travelling fairs was held in Albany in 1889.⁴

By late summer of 1890 everything was ready for the big event. Only no one bothered to tell the weatherman.

OPENING DAY

On Opening Day, Wednesday, September 11, things went swimmingly, according to *The Standard*. It was a cold 60 degrees and pouring. A month's rainfall came down in 36 hours.

The ducks, according to the reporter, were happy, but "the rest of the poultry hid themselves under their wings, poor things, and in the most complacent frame of mind possible, viewed the desolating downpour in company with the officers and managers of the vast enterprise as a visitation of providence."

A few city slickers from downtown Syracuse who didn't have much knowledge of the condition of farm lands after a two-day rain ventured to the fair in their business suits. "It was not uncommon to see a well-dressed man standing on one foot in a mudhole while, with his umbrella, he tried to fish out an overshoe from the sticky ooze where it had been sucked from the other foot."

Non-exhibitors didn't allow the rain to escape without notice. An ad by W.P. Butler's at 229-231 East Genesee Street, headlined: "No Rubber Suits and Lifeboats Needed to See the Best and Most Complete Line of Stoves, Ranges and Furnaces." The weather touched some raw nerves.⁵ An editorial the next day admonished visitors to "Take us as you find us."

The following day it was still raining until early afternoon, when the sun finally broke through. By 3 p.m. some 3,000 persons were on the grounds to see thousands of animals, a State Fish Commission display in the Horticulture Hall with live game fish swimming in a huge tank, and a family tree made of human hair from more than 150 persons . . . all for 50 cents.⁶

N.H. Howell of East Islip tells of the experience of an uncle, a gentleman named Andrewson, who wanted to bring something special to the fair.

Andrewson was a commercial fisherman who chanced upon a beached whale along the New Jersey coast. He saw the opportunity to make a fortune, and quickly put together a corporation which bought 60 barrels of embalming fluid. Next came the task of loading the bulk onto a barge where it could be immersed in the liquid. It was brought up to a reasonably sanitary condition, so to speak. From New Jersey the whale of a show journeyed up and down the waterways of the northeast, eventually arriving in Syracuse just in time for the fair, which was located within several hundred yards of the canal. The odor, after a summer of travel, was not in keeping with the fair's decorum. But the whale became a nearby sideshow, doing a "roaring business," according to Howell.⁷

While the first permanent New York State Fair in a Solvay pasture failed to break even financially and could hardly have been described as a great success, it was a beginning and its roots were to prove exceedingly strong, surviving wars, pestilence and perhaps most serious of all, periods of lack of official interest.

The element of permanency for the state fair meant a workable transportation system to haul the thousands of exhibits and shows; a reservoir of specialists to set up and decorate the demonstrations and to groom the grounds, and lodgings, restaurants and other accomodations for the tens of thousands of visitors attracted each day. This the Syracuse community could supply.

By 1892 the state fair was well settled into its new home.

The half-mile race track was a popular attraction for both the trotting events which included a purse of $300, and the four-horse team chariot races staged by Mr. and Mrs. A.A. McDonald of Chicago.

Japanese fireworks attracted a large evening crowd on Friday, with Governor Roswell P. Flower addressing fairgoers on Saturday.

After a day off on the Sabbath, the fair resumed on Monday with stock judging while the Honorable J.H. Brigham, Worthy Master of the National Grange, spoke on Grangers' Day.

Rail transportation via the D.L.&W. was every half hour commencing from 6:45 a.m. to 6:45 p.m. with fares at 10 cents one way and 15 cents round trip.[8]

And as an investment in future support, a children's day was inaugurated, with those under 16 admitted free.

THE CASE OF THE RICH ROOSTER

The 1893 fair attracted 8,000 entries, including the occupants of 16 cattle cars from Chicago. It also featured a voracious chicken.

A young woman from South Manlius was inspecting the rooster exhibit when she bent over the cage to get a better look at the strutting white leghorn. With a quick motion, he snapped at her ear. Down the gullet went a diamond earring. The lady screamed. The owner rushed over. As he explained later, "to decapitate the rooster (who serviced a large harem) for a mere diamond or two, would be like killing the goose that laid the golden egg." So he just bought the other stone too, and fed it to the bird the following noon before an appreciative crowd attracted by appropriate advertising.[9]

The gustatory appeal of the fair, in addition to diamonds, was highlighted by the traditional fight between the oyster vendors and the new-fangled frankfurter salesmen as to which was favorite. Bidding for state fair concession privileges included those for confectionery and nuts; bananas; oysters; cider; bottled drinks; tobacco and cigars; peaches; Coney Island sausages; grapes and popcorn. Others opened shops for sale of eyeglasses, decorated china, cement and glove cleaners, wire jewelry, glass blowing and engraving and nickel-in-the-slot machines.[10]

A perennial favorite even in those days was the Syracuse specialty, the salty potato, which is ascribed by some as the invention of a Syracuse barkeep, but which more likely was developed in the city's salt industry where workers servicing the huge kettles would drop the tiny potatoes into boiling brine. Once a heavy frosting of salt coated the outer skin of the spud, it would be split open, then treated to a thick dollop of butter.

GRIDIRON ACTION

The major weekday attraction in 1893 took place on Thursday afternoon when Syracuse University's football team met its then-arch rival, the Syracuse Athletic Association, in the Orange's first game.

The Syracuse A.A. was no pushover. In 1892 the club crushed the University in three straight matches following a 0-0 tie in the opening encounter.

Several thousand fans, with supporters evenly divided among those wearing an orange ribbon and others with the red and white of the S.A.A., filled the race track grandstand.

The field was leveled and covered with soft sod for this first (and last) football game there.

Most of the University players had just returned from vacation, while the club team had already held several weeks of practice. Besides that, the older S.A.A. participants outweighed the students by so much that "when they hurl their beef and brawn into the varsity line they mow them down like cornstalks."

In one play, Henry Hughes, an S.S.A. guard, picked up Syracuse University's halfback, a player by the name of Ulysses G. Warren, and carried him back 20 yards toward his own goal.

The result of all this was not surprising — a 22-0 drubbing for the University. It was the first of three humiliations that the Association was to inflict on Syracuse in 1893.[11, 12]

CONTINUED GROWTH

Later fairs demonstrated the need for continued growth if the concept of a permanent fair was to succeed. The 2,000-seat grandstand was replaced by a larger covered unit in 1897, while other permanent structures included buildings for livestock, horticultural and general farming exhibits. A forerunner of the present Home Center, a Domestic Building, was erected.

But through the "Gay Nineties" the fair was basically rustic. Enid Crawford Pierce remembered the earliest days when many of the exhibits were in tents.

Dirt roads and a few wooden sidewalks served as thoroughfares, while "three long horse barns were placed in a big triangle and snuggled nearby were some sheep pens and a pig barn or two."

She recalled the grassy paddocks for exhibiting the various horse classes during judging, and "where children cantered rather bumpily on their ponies, for the sake of their parents' pride in the resultant blue and red ribbons."

As befitted the pre-automobile age, if one wanted to get around one had to treat the horse as king. The horse barns and the people who frequented them were places and persons of wonder for the younger set.

"Everyone" travelled to the fair by horse and wagon — in carry-alls and lumber wagons, in buggies and in horse-drawn omnibuses.[13]

It was quite a sight, one oldtimer recalled, to see ten thousand horses hitched at various points in the grounds, each with its separate feed box and water bucket.

Grace S. Raines' memory is that some horses stood better tied than others. A fence with a single rail ran around the racetrack, so visitors could sit in the wagon and watch the trots (no one called them races in those days, Mrs. Raines maintains).

She remembered, too, that thoughtless roughnecks would sometimes lash out at the horses to see them jump, causing serious accidents.

Food was as much on the mind of the fairgoer of yesteryear as it is today, although a great deal of the preparation fell to mother. The night before the journey to the grounds, the farmer's wife packed the big basket of food.

Mouthwatering sections of chicken or boiled ham; a huge pan of baked beans smothered in molasses with chunks of salt pork peeking through the crust; loaves of homemade bread, the top just the right golden brown; fresh pickles; churned butter and an assortment of pies, cakes or cookies kept the family within sight of the wagon throughout the day.

That didn't mean that those who came without were in danger of starvation. In the days before tight health controls, almost any church group, civic club or organization could set up a tent, prepare home cooked and baked delicacies, and be sure of a crowd plus a healthy profit for the club.

Mrs. Raines was impressed by the candy butcher who produced "pull" candy, white if made of sugar, buff-colored if from molasses. The candymakers performed their art in an open tent, using muscle to stretch the mass of sweet-smelling fondant.

Instead of today's colas, which had not yet found their way into a bottle, lemonade and ginger beer were the big sellers, the latter in stone jugs.

Brightly painted horse-drawn farm machinery in reds, yellows and greens, with woodwork shined to a high gloss, filled sections of the grounds, while snorting steam traction engines pulled about huge threshing machines under a heavy pall of black smoke.[14]

Fireworks sprinkled the evening sky with multi-colored panoramas such as "Pain's Marvelous Spectacle," depicting Dewey's victory at Manila and the destruction of Cervera's Fleet in 1898.[15]

THE STATE TAKES OVER

The Syracuse show during this first decade was a workmanlike operation. But it was also costly, and its financing created problems as first a new grandstand, then the larger and better exhibition buildings had to replace outmoded structures. By 1900 the Society was in debt to the tune of $20,000.

Ten years of experience finally convinced the leadership of the State Agricultural Society that if the fair was to continue on a sound footing in Syracuse, the state had to take over.

The opportunity came suddenly. Lieutenant Governor Timothy L. Woodruff was elected president of the State Society in May of 1899, filling the vacancy created by the death of ex-Governor Roswell Flower.[16]

Governor Theodore Roosevelt and Woodruff both attended the 1899 fair and met at length at the Yates House with Society executive committee members, prominent citizens from Syracuse and several legislators. "Many hours were spent on this occasion in an earnest and most satisfactory interchange of opinions as to why the State Fair had lapsed into its present condition and as to what ought to be done," it was reported.[17]

They agreed to a state takeover. The legislature that spring had already passed a contingency bill acquiring the fair grounds and its buildings for $35,000 on the condition that the Society transfer to the state "all the right, title and interest to all its lands."

Former Governor Flower, as one of his last acts, had presided over this meeting and agreed to the stipulations.[18]

Governor Roosevelt tenaciously took on the task, recommending in his 1900 message to the legislature, that the state assume the fair's management.

Woodruff applied the pressure from the other side of the pincer, using his position as Society president to convince any diehards within the group they were in trouble without state control. He held good cards. For instance, the entire gate receipts for the 1899 fair were a mere $3,000 more than the revenue from admissions to the 1850 state fair and even less than several recent county fairs had yielded.[19]

Upon paying a visit to the fairgrounds, Woodruff remarked, "I was disappointed to find stubble fields where green lawns should be; paths of cinder instead of a less disagreeable material; a total absence of shade trees and even ornamental shrubbery, no floral decorations of the grounds worthy of consideration, and all the buildings sorely in need of a coat of paint for the purpose of preservation and attractiveness. The race track on this, the State Fair grounds, to my surprise, was only a half mile track, and the inclosure and paddock far from inviting to the eye or the comfort of the spectator.

"It was perfectly apparent why these conditions existed — the managers, who appreciated more than anyone else, the necessities of the care, instead of having a surplus out of which to defray the expense of correcting these conditions and making necessary improvements, as a matter of fact had to face a big deficit every year."[20]

Shortly afterward the bill was on the governor's desk for signing.

This law had another tooth. The debilitating competition for exhibitors and spectators resulting from county shows scheduled during state fair time, was combatted by threatening to eliminate their share of state appropriations. Competition from this corner immediately dropped off.

TURN OF THE CENTURY

As the new century began, the state fair once again shifted gears.

An initial step under state control was to improve the racing plant and increase revenue from the horse races, which were long recognized by the state as a bounteous source of funds. In 1895 for example, racing at both state and local fairs provided $95,980.54, much of which went for prizes for improving breeds of horses, cattle and sheep. The following year the New York State Racing Commission was created.[21]

Yet, horse racing had, by a resolution of the executive board, adopted just before the death of Governor Flower, been eliminated from the 1899 fair program.

Woodruff stated, "I do not understand that any of the present officers of the society were or are opposed to properly conducted horse races. I certainly am not, but, on the contrary, strongly favor the building of a mile track in order that we may have the fastest and most attractive trials of speed to be witnessed anywhere in the United States.[22]

THE SYRACUSE MILE

A sum of $10,000 was appropriated by the legislature in 1900 for construction of a mile track at the fair to accomodate the top racing attractions in the nation. By May 21, Contractor Frederick K. Baker had 15 teams and 40 men grading the area. Their deadline . . . August 1. They made it.

By all yardsticks, the following year's fair should have been a rousing success. The new track had had a year to break in; the Grand Circuit trotting races were booked with Abbott, the second best trotter in the world, scheduled for a record-breaking attempt. State officials had an extra 12 months to develop managerial knowhow.

It was also a time of major Upstate activity. President William McKinley was to visit the Pan American Exposition in Buffalo, while Labor Day parades guaranteed a busy early September. A horse show with 500 entries was being readied for the fair, as well.

The combination of Labor Day, Pan American Exposition and state fair provided an attractive menu for professional pickpockets from major Eastern cities who checked timetables to assure their presence at the best locations at the best times.

They were on hand in downtown Syracuse for Labor Day, where they left a visitor from Cincinnati minus "a green roll of $120," while he watched a parade in Hanover Square.

One lady lost her money when a thief slit open the bottom of her pocketbook which she'd tied to her belt for security.

Another woman was standing along the line of march when someone snipped off her silver-encrusted leather belt. She didn't know anything was amiss until she looked down to see her belt with the silver expertly skinned off, lying at her feet.[23]

McKINLEY

As workmen prepared the fairgrounds for the annual festivities, President McKinley and his wife were arriving in Buffalo from their home in Canton, Ohio.

It was only three days to the opening of the state fair when the president stood in a reception line in the Exposition's Temple of Music. It was late in the afternoon, and an organ recital had just concluded.

Among the visitors lining up to shake the chief executive's hand was a non-descript, youngish man who's only unusual feature was a handkerchief covering his hand. The time was 4:07. A Bach sonata furnished a classical background.

The man moved along until he was scarcely two feet from the president. McKinley smiled and offered his hand. The man knocked it aside and fired a pistol concealed under the bandage.

The first bullet bounced off the president's breastbone, the second entered into his abdomen.

It was later disclosed the assailant was 28-year-old Leon Czolgosz, an anarchist who was from Detroit, Chicago and other major cities.[24] But though he was obviously seriously wounded, the president called out, "Don't let them hurt him." Czolgosz was heard to mutter, "I done my duty."[25]

Over the long weekend the president appeared to rally even though gangrenous blood poisoning was running through his body. Although New York's Governor Benjamin B. Odell hastily cancelled plans to visit the state fair, the opening day went off as scheduled on Monday, September 9 under sunny skies.

Even though the great Abbott failed to show up, 10,000 visitors crowded the grandstand and railing around the track to enjoy the races. The two steam railroads and the Lakeside trolley serving the grounds were filled with fairbound travelers throughout the day.

Rains hampered fair activities the following few days, but headlines indicated the president had passed the crisis stage. The nation collectively breathed a sigh of relief.

The weather interfered with another epoch-making event. A few days earlier more than

100 automobiles left New York City on a 500-mile endurance race to Buffalo over what at the time passed for roads, with Syracuse and the fair an important stop.

The mud-spattered racers, by then winnowed down to 55, arrived from Herkimer, stopping in front of Syracuse's Yates Hotel for the evening, with David Wolfe Bishop in the lead. He was to give a one-mile speed exhibition around the fair oval, but the soggy conditions of the track resulted in cancellation of the visit. So the drivers, guests of the Syracuse Automobile Club, left the following morning for Rochester without a fair stop.[26]

SOMETHING ROTTEN IN THE APPLE DISPLAY

A "scandal of sorts" rocked the apple exhibit when one of the judges accidentally discovered something was rotten in connection with a couple of the fruit displays. Apparently perfect specimens of Baldwins, Russets and other species were set out for show and suitably rewarded with ribbons.

As he was passing his hands incidentally over the surface of one of the fruits, the judge noticed a slight depression under the little sticker used to label the variety.

On checking further, he found the sticker covered a hole . . . and that someone was at home. The apple, which had been marked a "first," was disqualified.

Further investigation bore additional fruit, and it was discovered a number of other prize-winning apples by these exhibitors had to be thrown out — literally and figuratively.[27]

Elsewhere on the grounds decisions were made to continue the fair on Saturday because of the horrendous weather earlier in the week. The president had had somewhat of a relapse, but no one locally thought it more than a temporary setback.

"THE PRESIDENT IS DEAD"

Syracusans awoke to headlines.

The president had died at 2:15 a.m. Saturday, September 14, with the words, "Good-by all, good-by. It is God's way. His will, not ours, be done.[28]

The fair was finally over.

Early in the morning crews commenced the job of striking tents, packing exhibits and closing down booths.

The flags over the various building were lowered to half-mast. Workers went about their tasks quietly.

The treasurer of the fair, a Mr. Brown, reported attendance for the week would not exceed 56,000. As he paid out funds to workers and creditors, a large loss was anticipated.

One of the commissioners was heard to say, "We are all sick and want to get home as fast as we can."[29]

THEODORE ROOSEVELT AT THE FAIR

It was not in the nature of Americans, nor of the community, to remain in a state of depression for long.

After reaching its low point, the rest of the decade was on an upward beat for the fair. America was feeling its oats. The automobile was coming of age. President Theodore Roosevelt was carrying the big stick, and the rest of the country was joining him in enjoying life.

The 1903 state fair was conducted in a carnival atmosphere which extended to downtown as the new president agreed to visit the city for a giant parade of welcome, then a major speech at the fair.

It was a year when women were saluted in a very special way with the opening of the fair's Women's Building.

And it was the first year the upstart automobile was to take to the mile track after the disappointment of local enthusiasts when the 1901 500-mile endurance run by-passed the grounds.

Cars were, by 1903, here to stay. Among the exhibits at the fair were the Rambler Gasoline Car, the Waverly Electric and the Cadillac with the detachable tonneau, even the Iroquois, called by its manufacturer, "the most perfect automobile made in the world." Syracuse's locally produced Franklin car was making a name for itself, too.

For the ladies there was the new "auto coat", a dark brown garment extending to the shoe tops (and apparently matching the color of mud). With it was a matching cap, a cape, also attached, and goggles.

The festivities opened on Labor Day with four major events going on simultaneously . . . the annual parade downtown; the opening of the fair; the ribbon cutting for the Women's Building and Theodore Roosevelt's appearance in both the city and the fairgrounds.

Syracusans worked for weeks decorating the city's center with flags and bunting. Electric signs of varied colors signalled "welcome" to the visitors. Reviewing stands were constructed, and a complex series of activities were timetabled from the moment the president's railroad car arrived.

Huge throngs came by rail. The New York Central ran special trains carrying 10,441 people from Auburn, Rome, Watertown and Ogdensburg, Mohawk and areas west. The Lackawanna specials came from Binghamton and Oswego with 2,609 travelers in 37 cars.

Local hauls from jammed downtown terminals lugged 9,000 persons out, with the Lackawanna running steam trains every 15 minutes. The Lakeside trolley was said to have carried 40,000 persons on that one day.

The president arrived at the D.L.&W. station for his 13-hour visit at 9:30 a.m., embarking in a carriage drawn by a two-horse team. His trip to the fair was a tremendous success. Roosevelt spoke for 58 minutes before a crowd of 15,000 which had removed its collective hats as the president took his place on the rostrum.

The president warned against class divisions and appealed for steadfast faith in the "first principles of the Republic." The Reverend Father Michael Clune of St. John's Cathedral remarked in a masterpiece of overstatement, "Such words were never spoken since Lincoln."

It was noted the president was closely guarded by secret service men throughout his visit, several plainclothesmen walking alongside his carriage throughout the parade.

A reporter said, "It is practically a certainty that had an attempt been made to harm the president it would have resulted in a dismal failure. An illustration of the precautions they take in guarding the chief executive . . . is shown by the presence of two guards underneath the floor of the speaker's stand, who were placed there to see that no bombs or infernal machines were placed there by some fanatic."

Even so, the difficulty in protecting an energetic president was obvious. Hundreds of fairgoers milled around cheering wildly everywhere the president went.

Despite one incident when a man was "slightly trampled" by a horse, the president was safely on this way by 10:30 p.m.[30, 31]

OTHER ATTRACTIONS

The Women's Building was formally opened by State Fair Commissioner A.E. Perren, and accepted on behalf of the women of New York by Mrs. Max H. Schwartz, chairman of the Women's Committee.

Congressman M.E. Driscoll in his welcome to the women, said the commissioners were not only gallant, but wise when they surrendered to the popular call by women for a building of their own.

Lena Bentley, a violinist, consented to play the congressman's favorite song, "Rosie O'Grady."

The horse show, by now a popular annual fixture, attracted the usual coterie of society folk, with parking spaces filled by the drags and traps of the show's patrons.

The Faugh-a-Ballagh, the public coach running between the Yates Hotel and the fair ground, tooled in to the merry sound of the coach horn, with Dr. John L. Wentz "on the cushion." Among the vehicle's happy party of 15 were Mr. and Mrs. J. Campbell of New York; Mr. and Mrs. Willaim B. Gere; the Misses Comstock; Mrs. H.R. Peck; Howlett Durston and Forman Wilkinson.

Upon arrival the party debarked from the carriage and entered the boxes, while the coach then moved onto the tanbark as an entry in the road class event.

It was noted that Thomas A. Maitland was one of the heavy harness judges. "Tommy" is just back from abroad, the story continued, "and wore a remarkable pair of spats and fetching yellow gloves, which set off a blue suit in striking manner."

Elsewhere as well, the 1903 fair proved satisfying to visitors.

The smallest locomotive in the world pulled a trainload of men, women and children over a half-mile track for just five cents.

Not everything was tasteful.

Women in particular, were said to admire the exhibit of a two-headed baby. "Thousands are viewing this little wonder," with women "loud in their exclamations of wonderment as they gaze on this exhibit," which received the highest recommendations from the metropolitan press when it was shown at Coney Island.[32]

DAY OF THE AUTOMOBILE

The day of the self-propelled land vehicle was no longer to be denied at the race track. While it often took hours for the few passenger autos to wend their way through the state fair crowds from one end of the grounds to the other, the mile track was something else.

An eight-event card of auto races was put together by the sponsoring Syracuse Automobile Club. The incomparable Barney Oldfield and his car, the "Baby Bullet," was the major attraction.

He was to headline the show with Jules Sincholle of Paris, France and Henri Paige, a Frenchman who made New York City his American home. The two drivers from overseas were to appear in identical cars of the "Paris-Madrid Type," a Darracq racer which had averaged 60 miles an hour for several hundred miles before the Madrid race was stopped at Bordeaux by French authorities after a series of fatal accidents.

The membership of the Automobile Club of Utica agreed to form a parade for the 50-mile run on Friday night (perhaps more to help each other out along the rugged roads than to flabbergast onlookers), and to witness the Saturday events.

When the first cars lined up on the track, an excited throng of 7,000 was present, drawn by stories of the great European road races and the Vanderbilt Cup events on Long Island.[33]

Barney Oldfield however, was a no-show. His car, driven by a 22-year-old substitute, Frank Day, crashed during a record attempt at the Wisconsin State Fairgrounds. The "Baby Bullet" toppled over, killing the driver.

Henri Paige was not in attendance either, since he was recovering from injuries suffered earlier in a race in Detroit.

Still, the onlookers weren't to be disappointed. Dan Wurgis of Cleveland, Ohio, driving a stripped down Oldsmobile, set a new world's record for five miles for cars of under 1,200 pounds by turning in a time of 5 minutes 49 seconds. The previous record in the class had been set by Syracuse inventor John Wilkinson in a Franklin at the Empire City track in Yonkers (presently known as Yonkers Raceway).

The vehicle itself was a crowd pleaser, with the car simply an engine, chassis and small seat for the driver, who dashed along exposed to the elements and flying stones.

Later the same John Wilkinson, driving from the scratch position, won a five-mile handicap run for road cars from Central New York. He passed two contestants on the backstretch to win.

Sincholle took his Darracq out on the track and ran a solo lap in the then-incredible time of one minute 3/5 second for the mile.

He also captured the 10 mile race for cars weighing less than 1,800 pounds, running the 10 laps in 10:36 2/5 to defeat F. A. LaRoche, also in a Darracq and Dan Wurgis driving an Olds. A fourth car, a locally-built Franklin, was the sentimental favorite but failed to finish.

Winner of the race itself was a foregone conclusion by the end of the first mile when Sincholle crossed the line seven seconds ahead of LaRoche. He continued to widen the gap without extending himself, to finish with a half-lap lead.

Syracuse area spectators had a chance to cheer another favorite in the five mile race for Winton touring cars with tonneau attached. Albert E. Petrie, a Syracusan, was behind the wheel of George S. Larrabee's car. He kept on the tail of H. H. Mundy's auto as the latter led for the first few circuits. The pressure commenced to tell however, and going into the fourth lap the leader pushed too hard. The overheated oil caught fire and Petrie slipped past the flaming auto to go on to victory.[34]

THE "DIP DETECTORS"

State Fair time, as described earlier, was recognized by pickpockets as "easy pickings," and they descended on Syracuse from all directions.

As former policeman William O'Brian said, after the fair closed "we would usually find a wagonload of pocketbooks — empty — all over the place."

Pickpockets are specialist criminals, usually staying with this form of crime, and usually becoming known to police officers in various cities.

It was O'Brian's idea as acting chief during the 1905 fair that Syracuse import the best detectives from other cities, send them out on the streets and pick up every known or suspected pickpocket they came across.

Eight of the best "dip detectors" in police circles — from Boston, Chicago, Cleveland, Denver, New York and Toledo, were brought in, all expenses paid.

They were accompanied by 16 local patrolmen, promoted to plainclothes duty for the duration of the fair.

Going out in teams of three, the eight squads brought in 38 prisoners, mostly top pros who were out-of-towners in town specifically to work the fair crowds.

It was downright embarrassing to be picked up by hometown officers whom they'd thought were safely behind in Boston or Toledo, and it was doubly embarrassing to learn that when they beat the vagrant rap by flashing a big roll of bills they were being charged with common gambling, fugitive from justice or similar crimes.

They were arraigned, the cases adjourned for a week, and then the culprits were assigned to cells until the fair closed. The few who weren't caught took fright and left town.

The following year only 20 were picked up; a year later, just 10. Syracuse became known as an unhealthy spot.[35]

MYSTIQUE KREWE

While the fair's acceptance expanded, Syracuse's civic leadership felt that growth on a Solvay pasture wasn't enough. After sundown the fair left nothing for visitors but to go home. Hotel and restaurant operators and others concerned with the city's economy, including Mayor Alan C. Fobes, sought a downtown attraction mutually beneficial to the fair and the city.

The task fell to the Syracuse Chamber of Commerce, and in turn to the Chamber's entertainment committee headed by John W. Smith.

In 1904 a three-night series of parades was well received, but then came an idea which was a true novelty, at least in the northeast: the development of the Mystique Krewe and establishment of an annual night-time carnival to coincide with the fair.

Following its debut on September 12, 1905 the carnival quickly became a way of community life. Patterned after New Orleans' Mardi Gras, but without the pre-Lenten significance, the Syracuse project was called the Kanoona Karnival, a combination of Indian lore, history and community fun, all under bright electrical displays in downtown streets.

Kanoona (or Ka-Noo-No) was the name of the City and Bay of New York and, according to early local historians, thence applied to the state. The word was believed to be of Mohawk Indian derivation — from gannona, meaning a well-watered or guarded land. (Gannona — bottom of the water; Gannonna — to guard). With its abundance of canals, creeks and lakes, the Syracuse area was well suited to the name.[36]

The theme of this first Karnival was the Spectacle of Hiawatha, with 12 floats which proceeded past the grandstand erected in Clinton Square, the area lighted by hundreds of candles placed along the canal's banks.

Among these ambitious moving displays were some depicting the legendary combat between the Indian hero, Hiawatha, and the Great Mosquito; the meeting of Hiawatha and Minnehaha; the death of Hiawatha's daughter, crushed by a great white bird, and Hiawatha ascending in a white canoe, plus other floats of a more recent and historic nature.

A king, Daniel M. Edwards, and queen, Miss Elizabeth Evans, presided over the Karnival, with Mayor Fobes turning the keys of the city over to them for their week-long reign.[37]

The "Krewe" in "Mystique Krewe" represented the five Indian Nations, each having its chief and subchiefs, with rivalries developing between the Krewe nations as to the number of adoptions and the application of appropriate Indian names to the various prominent citizens so chosen.

The Onondaga Indians themselves were caught up in the enthusiasm, truly adopting some of these individuals into their Nation. Among the Ka-Noo-No chiefs so adopted were H.W. Smith, Harry H. Farmer, later to become a hero of the Great War and mayor of the city, and George N. Crouse.

By the end of the decade food stands abounded with decorations; Clinton Square had a circuslike atmosphere dominated by overhead lighting, huge bleachers and much noise and music.

The concept worked.

Instead of going directly to the fair, many tourists first made their way around the city to admire the decorations. Syracuse was becoming famous as "the only Northern City with an annual carnival approaching New Orleans in originality, pretentiousness and beauty."

Harpers Weekly commented, "Do not let them tell you that an American town cannot enter into the carnival spirit, and still preserve the graciousness and a certain underlying sense of decorum. Tell those scoffers to go to Syracuse during State Fair week. They will see a demonstration to the contrary. Salina Street alight with incandescent beauty, and with row upon row of eager citizens."[38]

As the fair became a nighttime activity in the early 'teens, the Karnival Spirit, having served its purpose, gradually faded away.

ARRIVAL OF THE AIR AGE

The air age, exclusive of the hot air balloons which had been a fair staple for many years, also came to Syracuse in 1905 with the arrival of Carl E. Myers and his airship. This huge air bag which held 7,000 cubic feet of gas, carried aloft a dirigible car and motor. A rudder enabled the pilot to steer.

Captain Myers drew large crowds as he navigated the vessel between the fair's buildings and up to a height of 600 feet before bringing it safely back to earth.[39]

Six thousand fans skipped work or school to view the so-called "whiz wagons," crowding the course for the 1905 automobile races, even though a heavy rain Sunday night almost resulted in their cancellation.

Glenn Curtiss of Hammondsport, destined to become one of America's greatest aviation pioneers, flashed to new, flying start, one- and two-mile records on his motorcycle.

Then Dan Wurgis clipped a fifth of a second from the mile record in his 30 horsepower "Rice Bird" car.

Barney Oldfield, the champion of at least the New World, finally did show up in his Green Dragon. He turned in the fastest five mile run anywhere for the year, with a 4:38 2/5 time for the five laps (the world's record at the time was 4:29, also by Oldfield).

The champ then contested Walter Winchester, a local driver, in a special two-out-of-three heat match race. Oldfield, always the showman, let Winchester lead until the last turn when he burst ahead for the win in the credible time of 5:22 ½.

As he flashed across the line, spectators were rewarded by a big grin from Oldfield, who's trademark cigar remained at the same cocky angle. The second heat was a replay, with Oldfield again nipping Winchester at the tape in 5:15.

It remained for M.T. Bernin, however, to provide the greatest thrill of the day in the five-mile national championship run. Driving a Renault, he laid off the pace for the first few laps, then opened up to tear past his two competitors, Guy Vaughn and Wurgis (Oldfield had not entered). As he charged into the fourth turn at close to 90 miles an hour, Bernin's car swerved and then slewed into the fence. Heavy rails, installed for the purpose of keeping horses on course, not high-powered autos, "shot into the air like straws."

Bernin regained control, but obviously in pain drove slowly to the wire as the others passed him. A half-dozen doctors ran from the crowd and hustled the driver to the fair's hospital where eight stitches were taken in a jagged cut over his eye.

When the bandaged driver was asked at dinner what happened, he said, "Oh, I was going too fast, I guess."[40]

FOR "A GREATER FAIR"

A *Post-Standard* editorial at the fair's conclusion congratulated the administration on developing the exposition to the point where it was almost self-sustaining despite a low admission fee and minimal annual appropriations, but stressed that unless all profits were plowed back into permanent buildings and other improvements, Syracuse's fair would always be second to the great agricultural shows of the midwest.

"There is need of a large domestic building. There is need for a hall for machinery so that manufacturers of agricultural implements need not present their exhibits outdoors. There is need of a new grandstand at the track, because if it is worthwhile to have horse races at the fair there should be adequate accomodations for the spectators . . . there is need of a new cattle shed, built after the western model, with a ring for display of stock and seats for spectators, replacing the line of ugly sheds now on the grounds. There is a need of resting places about the grounds. There is need of dining halls."

The article concluded that New York, as a leader in a dozen lines of farm products, plus its other advantages, "can afford to spend money as generously as do Ohio and Illinois and Wisconsin and Iowa for its state fair."[41]

As politicians saw it, New York's prestige was at stake. And more groups than Upstate farmers were concerned. Assemblyman James Oliver of New York City told attendees at the 1907 fair that he had been authorized by Charles F. Murphy to say that "Tammany (Hall) stands for a state fair in keeping with the agricultural interests of the Empire State," and that Tammany's Democrat legislative members were ready to approve a one million dollar appropriation for new construction.

Governor Charles Evans Hughes, appearing the next day at the fair, represented the Republicans in calling for "A greater New York State Fair . . . New York to be at the head of all."

Hughes viewed the then-current fair buildings as wholly unsuitable, with tents which leaked and blew down, interspersed with scarcely better wooden shacks.

A statewide architectural competition to design a modern environment for the fair was established. It was won by Green & Wicks, a Buffalo firm, which advised a total outlay of approximately $2 million would be needed to execute it (The design was eventually followed to completion in the late 1930's).

FIRST CONSTRUCTION PHASE

Construction on Phase One of the Green & Wicks program was completed on the massive, hanger-like "Manufacturers and Liberal Arts Building" on the northeast section of the grounds in 1908.

In keeping with the effects of the various world's fairs and expositions on the minds of architects, the structure was the anchor for a monumental series of buildings around a landscaped quadrangle extending south from the main gate. Across the way to the west were the State Institutions and Dairy Buildings, connected by a colonnade.

The next step was taken in 1909 when the legislature approved $278,000. Subsequent new construction included then-modern racing stables and a poultry building.

Four years later a new grounds plan was adopted, including the addition of 27 acres to the northeastern part of the grounds. The following year another $200,000 was put up to build a fireproof building for exhibiting and stabling horses.[42]

"SPECTACULAR AND SCIENTIFIC"

By 1910 the fair had embodied the spectacular and the scientific to what many felt was the detriment of the agricultural purposes of the affair, even though purses and premiums by 1909 totalled $75,000.

The auto races had become a standard fixture, with such personages as former president Theodore Roosevelt serving as honorary referees. In 1909 more than 35,000 spectators were present to watch Barney Oldfield, with 1,100 automobiles actually parked inside the mile track. Vice President James S. Sherman was among the viewers. Ralph DePalma, considered by some as the world's greatest driver, established numerous records. (He was to win 2,500 races in his career.)

The true aeroplane arrived at the fairgrounds as well. J. A. D. McCurdy attracted 8,000 persons to the grandstand and paddock as he flew his Curtiss bi-plane three times around the area at a height of 200 feet. It was termed "one of the greatest drawing cards ever."

The unusual took its customary stance on the midway. Sharing one booth were Baron Niev DeBaroy, the smallest perfect man on earth (24 pounds, 28 inches high at the age of 25 years), and his mother, Baroness Sidona, the Hungarian Bearded Lady, who before her marriage "was one of the handsomest women in Europe, and today, with the exception of a long silken Van Dyke beard which has grown since, managed to retain her former charms."[43]

A BOYHOOD LOOK

John C. DuVall, a youngster during the early days of the century, remembered the fair as the biggest outdoor interlude between Labor and Memorial Days, with a special appeal for kids, since opening day was always free to children and schools closed for the occasion.

He remembered the distinctive "taste" of the fair. Most of the area was unpaved, and the walks and parking areas were covered with cinders. When churned by thousands of feet into a fine black dust, it soon filled the air and settled on everything within range. Visitors breathed it, wore it and swallowed it "by the tongue," as it stuck to candied apples, home made taffy and hot dogs.

DuVall was especially fascinated by the steam tractors which chugged away as serious-minded salesmen demonstrated their use in powering threshing machines or silo fillers, while others snorted up steep inclines or ran around in circles showing off their maneuverability.

To DuVall, who was later to become a professional motorcycle racer and for many years "The Voice of Watkins Glen" for the Grand Prix, the highlight was the final Saturday which became by tradition "Automobile Race Day."

The boys would be at the grounds when the gates opened and head for the track, getting as close to the pits "as perseverance and the police would allow." There "we could actually smell Barney Oldfield's cigar, hear (Ralph) DePalma direct the tuning of his Simplex, and see Bob Burman oil the drive chains of his gigantic Blitzen-Benz."

When the red flag (red instead of green, meant "go" in those days) waved, "universal bedlam broke loose."

"Like the Noble Six Hundred," DuVall recalls, "not all the cars made it. Vehicles and their components were fragile. High-pressure canvas-carcassed tires blew out; poorly balanced, over-stressed engines literally blew up; drive chains snapped; wooden wheels collapsed, or things just flew apart."

After the race the youngsters combed the pits or track perimeter for whatever came from the cars . . . discarded spark plugs or broken valve springs were prized most, but any tangible token, even an oily rag if it could be identified with the race, could be an acceptable memento.[44]

A busy afternoon at the Fair in the early 90's.
   Syracuse Newspapers

Rare photo of football on the Fairground turf in the 1890's.
Onondaga Historical Association

Carriage entrance at the turn of the Century.
Canal Museum of Syracuse

Midway throngs around turn of the century.
New York State Fair

Carriages around perimeter of race track.
Canal Museum of Syracuse

A champion high jumper in the early 1900's.
New York State Fair

Hanover Square takes on festive air in 1903 as President Theodore Roosevelt's procession is soon to make its appearance. Photo is taken from Clinton Street Bridge, looking east past Clinton Square.
Onondaga Historical Association

President Theodore Roosevelt's carriage passes along East Jefferson Street while secret service men walk alongside the vehicle during 1903 Fair visit.
Onondaga Historical Association

Touring cars line up for combat before crowded grandstand.
J.A. Seitz Photo

Driver looks somewhat worried, and for good reason, since he has to ride this car at speeds close to 60 miles per hour.

J. A. Seitz Photo

Professor Meyer flew this daring contraption in controlled maneuvers at 600 feet over the Fairgrounds in 1905.

Onondaga Historical Association

Safety is last for these unhelmeted daredevils. The ambulance stands by to their left as the match race gets the go sign.

J. A. Seitz Photo

Bernin found his ride at the State Fair races in 1905 rather unpleasant after the Renault smashed through fencing, leaving the driver with a bad gash on his forehead.
Onondaga Historical Association

A front gate floral greeting to the "guv."
Canal Museum of Syracuse

Fairgoers climb aboard State Fair Special at old West Shore Railroad Station on Syracuse's near west side during early 1900's.

King Ka-Noo-No I, Daniel M. Edwards.
Canal Museum of Syracuse

*Queen Ka-Noo-No I,
Miss Mary Elizabeth Evans.*
Canal Museum of Syracuse

Ka-Noo-No Karnival.
Syracuse Newspapers

Incline tests capabilities of this steam tractor during 1906 Fair.
Canal Museum of Syracuse

Cattle on parade.
New York State Fair

Waterfowl exhibit is always popular with younger set.
Onondaga Public Library

Barney Oldfield leads Charles Hanna in a match race over the Syracuse mile.
J. A. Seitz Photo

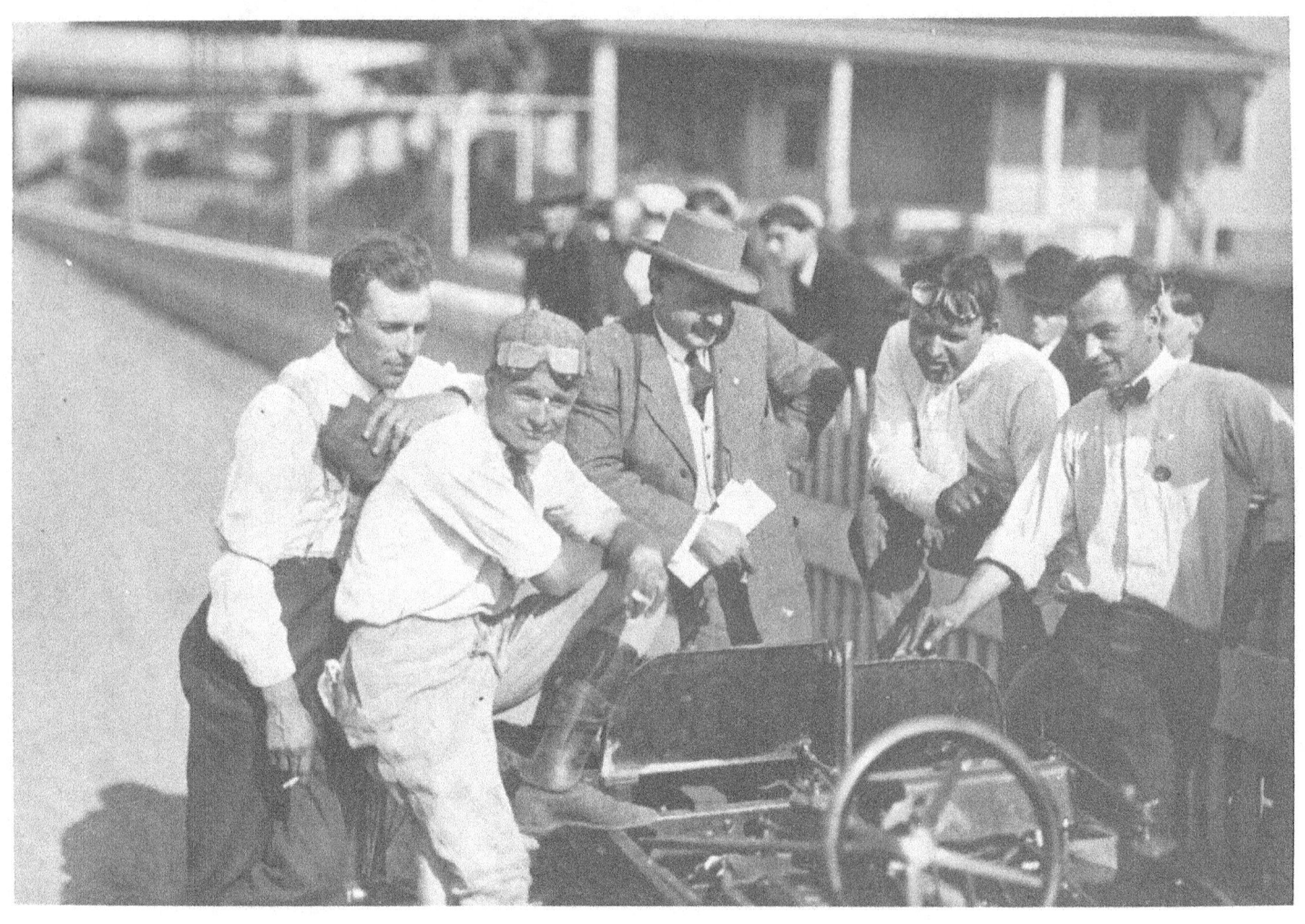

Barney Oldfield, with cigar, second from right, chats with fellow drivers and owners prior to a Fairgrounds race.
J.A. Seitz Photo

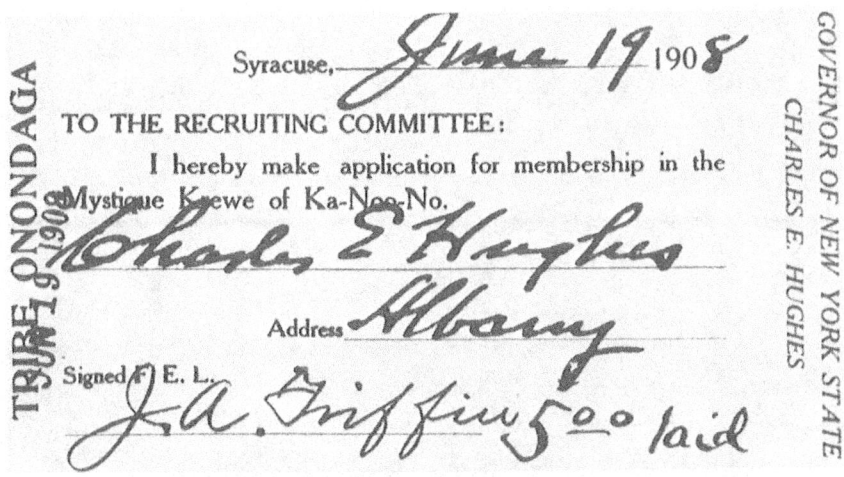

Charles Evans Hughes' membership application for the Mystique Krewe of the Ka-Noo-No.

Canal Museum of Syracuse

A 1908 float of Henry Hudson on the Half Moon.

Canal Museum of Syracuse

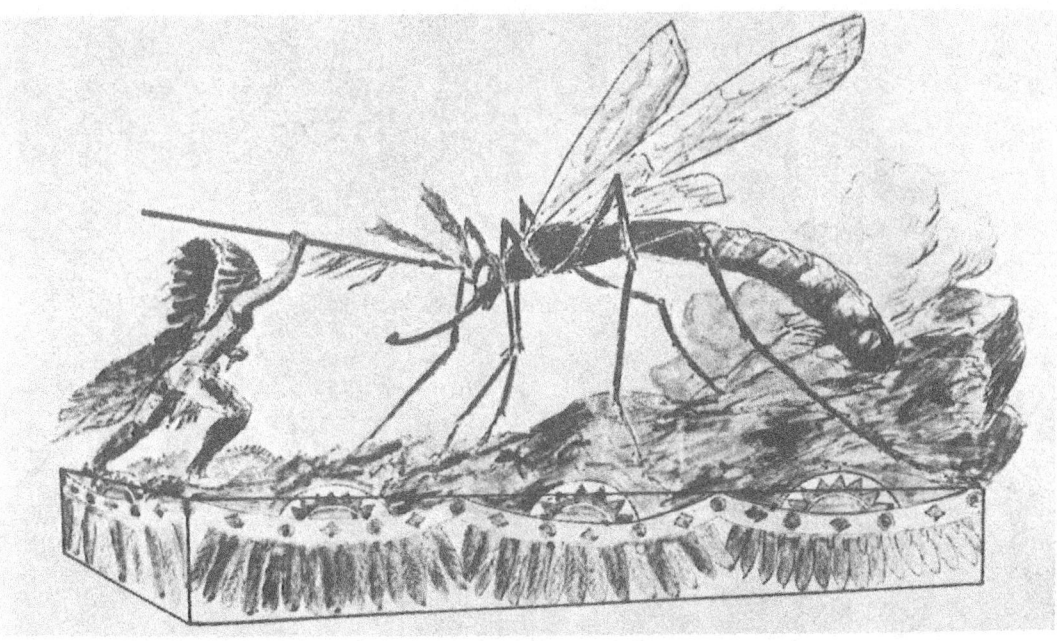

Indian legend of the Monster Mosquito is a float in the 1908 Karnival parade.

Canal Museum of Syracuse

Watching the 1908 Clinton Square Karnival are, numbered 1. F.R. Hazard; 2. State Fair Commissioner A.E. Perren; 3. Governor Charles E. Hughes; 4. Lt. Governor Lewis S. Chanler; 5. Past President Verbeck; 6. George N. Crouse, and 7. Secretary of State John S. Whalen.

Canal Museum of Syracuse

Clinton Square's Karnival spectacular in 1908.
Canal Museum of Syracuse

All lined up.
Onondaga Historical Association

State Fair Administration Building, circa 1910.
Photo by Harlow D. Curtis

State Fair staff strikes a pose.
Photo by Harlow D. Curtis

Alfred Gwyn Vanderbilt at the 1910 Fair.
J. A. Seitz Photo

The 2005-pound Percheron mare Gladys, owned by Charles M. Crouse, took first prize in its class in 1910.
Canal Museum of Syracuse

Governor Charles Evans Hughes addresses Fair audience.
Onondaga Historical Association

They're off . . . Past the judges' stand in 1911.
Canal Museum of Syracuse

J. A. McCurdy with his biplane in 1910.

Photo by Harlow D. Curtis

Ralph DePalma sets mile record in 1910.

Photo by Harlow D. Curtis

Syracuse Republicans greet President William Howard Taft as he alights from his railway car in 1911.

Onondaga Historical Association

Every space is jammed, including exhibit hall roof in 1911.
Photo by Harlow D. Curtis

Syracuse Automobile Club Racing Committee for 1910-11 racing at the State Fair are, from left, Forman Williamson, Howard C. Brown, Willet L. Brown, H.W. Smith, Arthur Benjamin and Fred K. Elliott.
J.A. Seitz Photo

Part of the infield crowd near the start-finish line for the 1911 auto races.

Syracuse Auto Club

Stunned spectators view fatal race car which killed 11 fans during 1911 races.

Onondaga Historical Association

Man on fence is believed to be Lee Oldfield, the driver who's car rammed through timbers in 1911 accident resulting in deaths of 11 spectators.

Syracuse Auto Club

Airplane attracts crowd in the infield.

Onondaga Historical Association

Fireworks and pageantry were aimed at building up nighttime attendance. Here Vesuvius erupts.
Onondaga Historical Association

Few could resist this PAIN fireworks poster.
Onondaga Public Library

A prize winning rooster.
Onondaga Public Library

Plows were still a basic way of life for the dirt farmer in 1912.

Dairy and Grange Buildings.
Canal Museum of Syracuse

CHAPTER III

The Days of the Pale Horse

America in 1910 was comfortably wrapped in a cocoon of prosperity and the rewards of 50 years of peace. Mass production was making goods and services available to everyman. Medical science was overcoming many of the great threats to life.

The Central New York community was enjoying its share of gemutlichkeit. The canal was on its downtrend as a way of transportation, but the automobile was becoming available to almost everyone.

The state fair entered the second decade on the upswing, with new buildings and a promising future. The auto and the airplane were standard fair features. It was an era of peace, complacency and optimism.

What could go wrong?

Fairviewers didn't have to wait long to find out. For others the days of reckoning were still a ways off.

The 1911 fair progressed favorably toward closing day. Saturday, as usual, was devoted to speeding cars. Officials meanwhile, were congratulating themselves on a successful show.

It was also the occasion for a visit by President William Howard Taft and his vice president, James S. Sherman. Following a 7 a.m. arrival, the president and his party breakfasted with community leaders downtown and then adjourned to the fair.

The race program was halted as President Taft and his entourage drove past the grandstand before a wildly cheering crowd. Then the first of two 50-lap races was run off. The winner, Mortimer Roberts, was timed in 50 minutes 53 seconds . . . almost a mile-a-minute clip.

By the time officials had lined up the second 50 miler, the president and his party had boarded the train and were well on their way.

Ralph DePalma was geared up for a fast run, and soon opened a substantial lead. Lee Oldfield, an experienced driver (although no relation to the more famous Barney), was behind the wheel of Number 11.

Even though the track had been wet down between races, the speeding cars soon churned up the dirt, casting the entire track under a surrealistic haze of dust and smoke made reddish by the setting sun. The stuttering roar of the unleashed engines and the squealing of tires and brakes as they skidded through the turns added to the aura. Those close to the track's surface were soon mesmerized by the fleeting blobs of color.

Witnesses said Oldfield held his own for the first few laps but engine trouble developed and by the 44th mile he was a lap behind.

A member of Oldfield's pit crew was watching as the "11" car moved down the grandstand straight on his 44th circuit. "He was going faster than DePalma and was pumping the gasoline with one hand while he steered with the other. Then, I am told, when he came to the first turn, instead of shutting off his engine and coasting around, he continued ahead.

"He swung out wide for the turn. It was near the quarter-pole bend as he was straightening out for the back stretch that the tire blew out."

In an interview obviously improved upon by a reporter, Oldfield said later from his hospital bed, "It was a nightmare . . . I was just straightening out for the backstretch run when I heard my tire burst. Then we went into the (fence) post and uprooted it. I shot into the air and seemed to thread my way through a maze of flying timber. I relaxed my muscles and fell limp. Then all was void.

"I looked about and there was a man with a newspaper over his face. I raised myself on my elbow, and the crowd seemed to surge around me. I sprang up and rushed over to the machine, where the engine was throbbing. I turned it off — I don't know how — then somebody took hold of my arm and led me away."

Part of the crowd of 50,000 was jammed into the hollow in back of the ridge upon which the infield fence had been built. That's where the car landed. By the time the carnage had been sorted out, seven persons had died and many others found injured, some critically.[1]

Such a person was William Sharkey of 214 Harrison Street, a 22 year-old who was placed with the dead on the fairgrounds grass. An intern, Doctor Robert J. McGuire, leaned over him, then turned to an attendant. "This fellow is still breathing."

Sharkey was taken to the Hospital of the Good Shepherd (now Upstate Medical Center).

By the next day four others had died, bringing the total to eleven, including a chemist, a school teacher, a bank clerk and an engraver and his nine-year-old son. But Sharkey was not among them. Three days after the accident he regained consciousness. Looking up, he saw for the first time a nurse, Elizabeth Rogers.[2]

Sharkey mended rapidly. He became a city patrolman on May 1, 1914; and in October, 1915 he and Miss Rogers were married.

Strangely, the accident failed to stop the race. Fred J. Wagner, the Automobile Association's starter, waved the yellow flag (then a stop signal) as DePalma came by, but he failed to see it, and completed the race in the world's record time of 47:21.65, soundly beating Barney Oldfield's old standard of 47:57 for 50 miles.

Lee Oldfield told investigators, "I never want to see a circular track and a white picket fence as long as I live."[3]

And for almost a decade auto racing was a non-subject at the fair.

WHEN THE SENATORS PLAYED THE REDS

The 72nd Annual Fair opened on September 9, 1912 for its six-day stand, with aeroplane races and the Grand Circuit bringing the fastest trotters and pacers to the grounds. There was much more.

The Washington Senators baseball team, with the incomparable Walter Johnson (who was to win 414 games during his major league career) were scheduled to play an exhibition game against the Cincinnati Reds on a specially prepared diamond in the infield.

A monumental pageant and fireworks display depicting the Last Days of Pompeii was staged, and Woodrow Wilson, the governor of New Jersey and winner of the Democratic nomination for president earlier in the year, appeared as a featured speaker.

The ball teams showed up as planned, but the field left a lot to be desired. One look at the pitcher's mound and Manager Clark Griffith of the Senators decided that his star pitcher wasn't going to play.

Even so, the Senators started most of their regulars, although the Reds went with a number of substitutes.

Without a grandstand or bleachers, the fans crowded along the foul lines and spilled over into the outfield. The bumpy turf led both Griffith and Hank O'Day, manager of the Reds, to caution their men not to take any chances when running.

Washington was off to an early lead and was ahead 3-0 after the second inning. Then, in the Red's half of the third, with the bases loaded, "Doc" Hoblitzel hit an easy grounder to pitcher "Long Tom" Hughes who muffed the play, letting two runs in.

Andy Kyle came home on Mike Mitchell's three base hit. He in turn scored on a single by Art Phelan. "Harvard Eddie" Grant, Pete Knisely and Henry Severeid followed with hits to bring in the last of seven runs.

The fans enjoyed the game though, a number marvelling at the speed of the major leaguers, with many baserunners scoring from second on singles. They enjoyed the umpire baiting of Washington's coach, Nick Altrock, baseball's clown of the era. And they enjoyed the speed of the game which was over in a merciful one hour 37 minutes, with the final score Cincinnati 14, Washington 5.[4,5]

THE FIRST AIR RACE

The weather was ideal for the first aviation race meet ever held in Central New York.

Showing up for the 10-mile aeroplane race were Beckwith Havens, an army pilot and William B. Hemstrought, both flying Curtiss racing biplanes, while Walter E. Johnson, a former Syracuse University athlete, flew a Thomas Racer.

Close to 10,000 persons filled the stands as the planes were tuning up.

Johnson was an immediate favorite. The fairgoers were entranced with his flyer "gracefully gliding from earth in broad, sweeping curves." The Syracuse pilot almost came to grief, however, as his landing approach was blocked by spectators running to get a better view. He had cut his engine and couldn't get enough power to ascend.

At the verge of stalling out, he kept aloft just long enough to miss the crowd. The machine went further upfield than planned. When the wheels finally touched down, Johnson skimmed the craft along directly toward the race track fencing.

It seemed impossible to avoid a crash, but by collapsing his wheels and dropping the plane onto its skids, Johnson succeeded in turning at almost right angles. The craft came to a stop a scant 20 feet from the rail

Finally it was time for the feature 10-mile race, 10 times over the circumference of the one-mile track. Johnson found the low altitude and the handling of his Thomas plane a perfect blend and easily defeated Havens and Hemstrought in the three-plane race, running off the event in 10:42, far slower than the times of racing cars in previous years.[6]

POMPEII DESTROYED — AGAIN

The fair resorted to night-time entertainment with special train service to order to bolster the gate and reach new audiences of working men and women.

It was obvious something of a major magnitude was needed. It was equally obvious that if Pompeii could again be destroyed by Mount Vesuvius' eruption, this time right in front of the grandstand, a lot of people would pay to see it. Such an event just happened to be in the repertoire of the Pain Fireworks Display Company of America.

A contemporary description enthuses:

"So vast and so all-embracing is their scope that the spectator involuntarily wonders at and admires the ingenuity and daring of the man who conceived them and the master hand that made their production possible."

"Pompeii is visited in her greatest splendor, with her citizens thronging the streets and preparing for the holidays. The first hour and a half is one of revelry, while Vesuvius rumbles sullenly, unheeded by the merrymakers.

"The revelers demand as the day closes, the Christians be killed. Just as the evil deeds are to take place, a blinding flash is seen for miles accompanied by a tremendous explosion 'that makes the very earth tremble,' as the top of the great volcano is blown off and fiery streams of molten lava pour down on the devastated city.

"Buildings and spires wave to and fro, then crash. Chaos reigns . . . the spectators fear for the safety of the hundreds of actors . . ."

Then . . . a brilliant display of Pain's Manhattan Beach fireworks of 66 set pieces, lasting nearly an hour.

The finale, as Conway's famous band plays "The Star Spangled Banner," is Old Glory waving in lines of fire.[7]

Other new fair wrinkles included improvements in the Women's Building which provided cool rest areas with rocking chairs for the comfort of the elderly, while "preambulators, cradles and little beds under watchful eyes of careful nurses and attendants are for baby or child who feels the 'sand man' at its eyes" . . . all without charge.

THE MID-TEENS

According to Roy E. Fairman of *The Herald-Journal*, long the dean of state fair newspaper reporters, in 1915 most visitors still reached the fairgrounds either by horse-drawn vehicles or the busy shuttle trains despite the emphasis on the automobile.

Organization of the state police as the fair's security arm was still two years away, and the grounds were patrolled by special police hired for the occasion. No uniforms were available, so they were equipped with the tall helmets in vogue at the time, and nightsticks. Sizes varied greatly, according to Fairman, with the shortest cop appearing little taller than his helmet.

Fairman recalled a phenomena of the pre-World War I era. Men and women, instead of wearing comfortable outing clothes, dressed in their Sunday best. Men sweltered on the hottest days in high, tight, stiffly starched collars, detachable cuffs and tight vests and coats. Women were garbed in gowns whose skirts reached almost to the ground and were so small in circumference that slashes were cut in them to allow the wearer to walk upstairs. He remembered as well there were few pregnant women in evidence, most of them remaining in relative seclusion during this period of true "confinement."[8]

THE CHILD PLAGUE

It slipped in on the hazy, pleasant days of August, following perhaps an afternoon at the lake or the amusement park. Or after a day of shopping for new school clothes in downtown Syracuse, and a stop for an ice cream soda on the way home. It was no respecter of locale. Its presence was felt in Skaneateles, then Jordan, Manlius and Eastwood, Fayetteville and, of course, Syracuse.

The year was 1916. The Great War in Europe which was to have lasted a few weeks was in its third year. The great battles of Verdun and the Somme were winding down. At the fairgrounds New Yorkers were preparing for another peacetime show.

The Allied Clubs of Syracuse had promised themselves there would be 50,000 people at the fair on opening day!

After several years during which night-time entertainment failed to substantially increase fair attendance, it was planned to reinstitute the Karnival.

Then little Clara, a suburban girl just eight years old, complained she didn't feel well. The doctor was called, even though it was Sunday night.

By Thursday morning Clara was dead. Diagnosis: Infantile paralysis, the summer killer.

Another girl, this time age 10, was stricken on a Wednesday morning. The doctor came that afternoon. Thursday the signs of paralysis became evident. That evening the doctor again visited the home. By 7:30 p.m. she too had died.

A local village was posted to warn off visitors. No children under 16 were allowed to enter.

Youngsters died in other communities.

Another day, more notices: 13 cases, two dead; school openings were postponed until late September for the Onondaga Academy and schools in Fabius, Pompey and Dewitt.

Headlines read: "Child Plague Kills Two More; 11 Other Cases."

Meanwhile, plans for the Ka-Noo-No Karnival continued.

All churches in Skaneateles closed down, and all children under 16 were quarantined to their homes as state health officials came in from Albany.

On Monday nine more persons were stricken. One died. The following day two more passed away. A doctors' conference was held to decide whether children would have a part in the Karnival.

State Fair Boulevard was opened by Lieutenant Governor Edward Schoeneck, offering a new, rapid way to the fair by car. Ironically, the new road passed just west of the Women's and Children's Hospital, one of the outposts against the dread disease.

On September 6 only two new cases were reported in Syracuse, and encouraging intelligence from Long Island and the Lower Hudson Valley indicated the epidemic was waning. Doctor William D. May of the State Health Department said a few days later, "The outlook for Syracuse is favorable."[9, 10]

When the 76th Fair opened, local residents were looking for some relief from the strain. A Monday crowd of 5,000 at Empire Court heard Charles Evans Hughes, the Republican candidate for president, criticize the administration's Mexican border policy, as he pleaded for one flag, one country and one national ambition.

He further called for arbitration to settle all disputes as he launched a heavy barrage against President Wilson.

An automobile parade of 3,000 decorated cars under clear blue skies took an hour to pass the Clinton Square stands.

Although the Syracuse Allied Service Clubs had previously forecast attendance for opening day at 50,000, under the circumstances the community's planners were pleased with the so-called "Sac-Bustr Day" turnout of 20,903 visitors. It was termed "an elegant testimony to the willingness of Syracuse to support the fair — an answer to critics."

The auto parade was "the longest and most picturesque daylight event ever seen in the city."

This infatuation with the automobile carried on to the fairgrounds as well, with one display including a complete car made from farm products.

The skin was of egg plants.

The radiator comprised corn and cucumbers.

Red peppers were formed into the crank.

The steering wheel consisted of traverse sections of beets.

Wheels were carved from scalloped squash.

Corn husks constituted the top of the car.

A lady was riding in the vehicle, her hair of corn silk, her hat of red peppers.[11]

The race track attraction featured a special exhibition of Directum J., the world champion pacer, driven by Thomas Murphy of Poughkeepsie, the horse shattering all existing records for a half-mile with a time of 55¾ seconds as against the 56 set by Dan Patch in 1903.

DeLloyd Thompson, the King of Aviators, took over the infield for a series of demonstrations including precision bombing and a loop-the-loop. The pilot had to be lifted into his plane because of a crushed ankle, injured in a crash from 400 feet in an earlier exhibition. During his Syracuse stay, Thompson's meals were served to him while he remained in bed with his left leg in splints and bandages.

Thompson endeavored to perform his loop the following day, taking off in windy weather. He reached 1,000 feet, the plane heading west. In two minutes he was beyond the horse show stables. Turning, he commenced his return trip with the wind at his back. His airspeed was 110 miles an hour. Instead of a question of looping, it was a gamble as to whether he could even get back.

Banking into the wind for a landing, Thompson could make no headway. In order to gain ground, he had to point the nose down. Then he flew into a low, scuddy stormcloud. The motor stopped. Spectators saw the plane for a moment as it hurtled toward the ground through the shifting mist. The motor stuttered, then came to life.

Thompson pulled out of the dive not more than 50 feet over the field.

Although the fair, which attracted 122,842 during its stand did serve as a diversion, the paralysis continued. The event, always a great attraction for children, was off limits to those under 16.

Midweek reports on the epidemic indicated another death with four more cases listed. Similar news was printed the next day. Two children died within a short time of one another in Jordan.

Then, on the 15th, only one case was announced. The seasonal disease had finally peaked, then abruptly dropped off, too late to help the fair.

THE FAIR BECOMES AN ARMY CAMP

Submarine warfare in the Atlantic and the pressures of the British and French for American intervention in the Great War finally broke down political resistance and in April of 1917 the United States was at war.

The fairgrounds soon took its place as a major army installation. Three days before Congress' approval of the declaration, Army Captain A.M. Pope of The Manlius School declared at the annual meeting of the Syracuse Chamber of Commerce that no better site existed for a training camp than the fairgrounds. The idea was adopted.[12]

A few days later, on April 12, a Major Howard K. Brown was designated to command a remount company being organized for fairgrounds duty. Thus, the first chore of a fairgrounds military unit was to serve as a receiving station for horses and mules.

Some 10,000 horses (heavy draft animals for the field artillery and lighter horses and mules for ambulances, pack and ammunition trains for the national guard) were assembled there.

On May 14, 1917, President Wilson designated the area as a "concentration camp." It became the recruiting center for the 38th, 39th, 48th, 49th and 50th Infantry Regiments and the 15th Field Artillery Regiment.

The Wall Street Journal commented editorially on Syracuse's choice: "ever a city of patriotism, pluck and persistence," and "whatever it has undertaken, it has carried through."

The newly designated Camp Syracuse was actually an amalgam of three individual camps. First was the fairgrounds; the second was three miles to the west, while the third was established between the two.

By July 17,000 soldiers were undergoing processing and training, with the first arriving on May 22; five sections of the seasoned Fighting Ninth Infantry from San Antonio, fresh from border activities in Mexico. These units were broken into nuclei for the formation of new regiments.

The five-mile stretch of grounds included 600 acres of pyramidal tents.

In the fairgrounds, cattle and livestock buildings became barracks, a cot to a stall, with many of the name plates of former occupants (cattle and horses) still remaining. Often these names became attached to the human incumbents, and were carried with them to France and later life.

Beds were installed in the Horticulture Building; tents under the grandstand, and Pine Plains was turned into a firing range.[13]

THE 1917 FAIR

Despite those who held that a fair could not be staged that year, the 1917 state fair opened on schedule on September 10, with the war and "production" the major themes.[14, 15]

The expansion camp, as one official remarked, "represents the mobilization of the country's manpower. The state fair will represent the mobilization of New York's food producing resources . . . On the fairgrounds this year there will be an intermingling of the armed defenders of the Republic and the men who, in the field and garden, in dairy and orchard, have been quietly toiling these many months to meet their country's critical needs."

On opening day Mayor John Purroy Mitchel of New York (who was to be killed later in the war in an airplane accident) discussed the high costs of living. Each day there were military drills, trench warfare in sham battles, barbed wire entanglement problems and other military demonstrations.

For three days the troops were on holiday to enjoy the fair and its displays. On September 13, Governor Charles E. Whitman received the soldiers.

As a wartime fair, the 1917 event was a marked success, proving the community's assimilation of the military into its fabric.[16]

It was not long after the fair that units began to move in secret out of the camp enroute to advanced training, maneuvers and overseas deployment. Although efforts were made to talk the government out of it, the camp closed down for the winter by November 1. Obviously, as Chase stressed, "Syracuse in winter is no place for an army camp."[17]

THE 1918 REOPENING

The camp was reopened in 1918 and soon regained its former prominence, although with a somewhat altered mission. In July, 11,000 limited service troops, physically qualified for many military jobs but no front line service, were ordered to the Syracuse Extension Camp for training. As it eventually turned out, many of them would have been safer living in no-man's land.

The fair itself became the War Fair, combining the peacetime agriculture and commerce with farming and industry in war, and further relating these to the combined effort of federal, state and county agencies.

A slogan was coined: "Fighting With Food."

Arrivals at the fair were greeted by a tremendous victory arch, rising 20 feet above the main entrance, with a span of 60 feet. It was dedicated to the men of the state who had already given their lives.

As a backdrop for the arch on the broad lawn of Empire Court was a display of the flags of England, France and America, worked in flowers along with the slogans, "Buy Liberty Bonds," Buy W.S.S. Stamps," "Produce Food," and "Conserve Food."[18]

Horse racing events still preoccupied many fairgoers.

Subways under the race track were completed while Pierre Lorillard of Tuxedo Park, one of the fair commissioners and a staunch supporter of horse racing, designed a genuine steeplechase course within the infield, complete with water jumps, hedges and other hazards, the races being set for closing day.

The fair itself was relegated to the inside pages of area newspapers, the war continuing to dominate the news. Field Marshall Haig's British forces moved on the northern flank of the western front while the Americans, many of them Syracusans with the Rainbow Division, were engaged in fierce fighting in the area around Chateau Thierry and the Marne. In the east, "Red Desperados" were "spreading terror and horror over Russia."

At the fair displays included highly popular War and Navy Department exhibits of models of the latest ships, guns and aircraft, with large collections of relics from the front. Demonstrations of Browning and Lewis machine guns vied with farm tractor tests for public attention.

The trotters again featured class talent, with Tommy Murphy in red and blue silks driving Directum J. to a sensational victory in the $2,000 Chamber of Commerce Pace in 2:06.

Although the community bravely continued the fair, it was obvious people had other things on their minds. Only 62,322 persons attended the 1918 affair![19]

OUR GREATEST CATASTROPHE

But the story of the camp was as yet imcomplete.

Within a week it was to be the epicenter for the greatest disaster ever to strike a community which had faced a devastating explosion at Split Rock a couple of miles from the fairgrounds which cost 50 lives just a few months earlier (July 2); had 10,000 of its sons and daughters serving in the military and which, during its long history, had experienced epidemics of malaria, smallpox, cholera and infantile paralysis.

In mid-September Syracuse homes were receiving news of the heaviest casualties of the war (although unlike the great drives earlier in the conflict, this time there was an air of finality about the Allied campaign).

Seemingly unrelated to any significant event, during the last days of the fair several recruits had reported on sick call. Their symptoms varied. The sickness came on suddenly. Headache, severe stomach problems, fever ranging from 101 to 105 degrees, general weakness and "aches all over." In from five to ten percent of the cases the illness degenerated into massive pneumonia. These symptoms seemed to mirror the characteristics experienced by soldiers in Camp Devens, Mass. on September 8.

Sufferers complained it felt like "being beaten all over with a club."[20]

Newspaper stories on the 19th revealed the Assistant Secretary of the Navy, Franklin D. Roosevelt, was recovering at the home of his mother following pneumonia which developed from what was described as a bout with the Spanish Influenza.

Another item out of Sweden related the death from flu of Prince Eric.

In some places the course this disease took were bizarre. One man in Australia recalled riding on a trolley. By the time he had arrived at his destination, three passengers had collapsed and died.[21]

As a safety measure, Dr. A.J. Gigger, Syracuse's bacteriologist was instructed to take blood specimens at the camp "in view of Spanish Influenza in some of the eastern camps." Shortly afterward a dispatch from Camp Devens revealed an ominous turn. Fifteen deaths were reported there.

On September 21, one doctor sought to comfort the community. "I think" he said, "it is simply the old Syracuse influenza caused by damp weather." The United States Public Health Service recommended "bed rest, good food, salts of quinine and aspirin."[22]

Syracuse hospitals were already overcrowded, with 200 military cases admitted.

By September 22 there were 500 from the fair area in the hospital; the camp was in strict quarantine. One soldier, Raymond Berkland, 22, of Norwood, Mass., a private, died at the Hospital of the Good Shepherd. He had been admitted on the 20th. Others died at nearby hospitals. Military deaths were soon averaging five a day.

The death rate among the soldiers continued to climb. Next it was eight in one day. Then 10. On the 27th, 12 more, with another 26 listed as critical.

On September 29 the camp reported 901 cases out of a population of 15,000 men. The fair's Poultry Building was filled with some 600 of the sick.

As military patients crowded every local hospital, the Red Cross sent 800 women who had taken its courses in first aid and home nursing into the hospitals. They relieved the hard-pressed nurses, several of whom had died. More than 100 were ill. Graduating nurses were called in.

Doctors gave up private practice and worked night and day in the wards. All elective surgery was postponed.

The draft was halted as the military tried to avoid bringing any more men into contact with the flu. At the same time it was noted that the young, healthy and vigorous seemed the most likely to fall prey.

By October 10, the camp's administration reported cases were declining, with 258 soldiers on the sick list. A total of 163 had died. It was apparent the worst was over, at least at the fairgrounds which had seen 4,000 cases during the epidemic.

But it wasn't over in the city. Some 8,000 cases struck the residents of Syracuse. The flu hit Winchell Hall at Syracuse University. The school's infirmary called on Red Cross volunteers to help with the hundreds of sick students. In Skaneateles the Packwood House (now the Sherwood Inn) with its 40 bedrooms became a temporary hospital.

Life went on. The Fourth Liberty Loan went over the top, with every whistle and siren in the city letting out a tremendous blast on Saturday evening, October 19. A loan parade in Chicago brought out the caution to follow the walk by "rubbing the body dry and taking a laxative."

But it was as though the whistles had been a special signal. On October 25 the disease had run its course to the point where the ban on public gatherings, including school, was lifted and the community slowly returned to normal.[24, 25]

Even during the worst of the ordeal the epidemic failed to take the front page away from the war. And now, after the death notices had been duly carried, the whole affair was figuratively swept under the rug.

The Spanish Lady left Syracuse and the fairgrounds. And, in due course, so did the military. The latter was not to return in force for another generation.

Syracuse, with a population of 161,404 had 1,052 deaths from September 28 through March 1919 which could be directly attributed to flu. The high was reached on October 19 when 253 died in the city.[26]

The decade was one best left behind.

Theodore Roosevelt draws applause during a fair visit.
Onondaga Historical Association

Cincinnati versus Washington in 1912 had overtones of sandlot game.
Onondaga Public Library

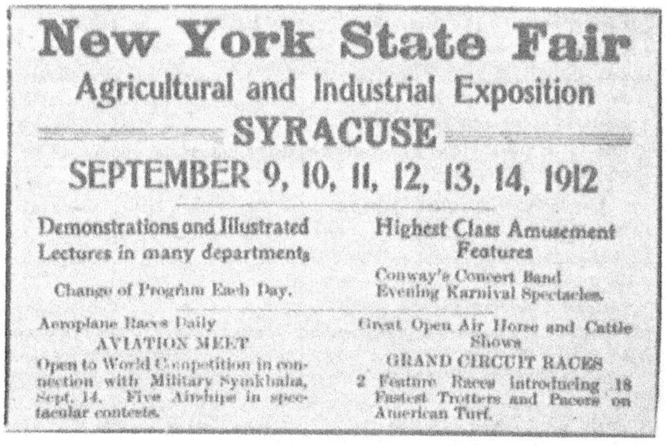

Ad in 1912 Lakeside Press helped bring out large crowds to aeroplane races, Karnival spectacles and the Grand Circuit Trotting events.
Onondaga Historical Association

Interior of Poultry Building with aquatic birds in foreground.
Canal Museum of Syracuse

Front straightaway looking east past grandstand.
Canal Museum of Syracuse

In the shade. A pleasant place during the heat of the day.
Canal Museum of Syracuse

Woodrow Wilson at the Fair during his election campaign in 1912.
Onondaga Public Library

The manufacturer of Queen Bess Pancake Flour managed to work in his product in this souvenir post card.
Canal Museum of Syracuse

See-saw time circa 1912, while parents enjoy the fair.

Onondaga Public Library

The Big Cheese — all 6,500 pounds of it — enroute to the 1913 State Fair.

First State Fair exhibit by then-fledgling State College of Forestry at Syracuse University, featured a 1913 demonstration on how to creosote fenceposts.

NYS College of Environmental Science and Forestry.

"Winter Strawberries" from Pulaski, N.Y. at 1914 State Fair.

Canal Museum of Syracuse

The 1913 Fair when "canvas" still prevailed at the annual show.
Onondaga Historical Association

Mounted police detachment in formation at the Fair.
Canal Museum of Syracuse

Detachment of New York City mounted police on hand for 1916 Fair.
New York State Fair

Food column dramatizes Empire Court area for 1916 Fair.
Onondaga Historical Association

Sac Bust'R Day to kick off 1916 Fair suffered from infantile paralysis epidemic.
Onondaga Historical Association

Sac Bust'R mail publicizes 1916 Fair.
Onondaga Historical Association

Old timers at '16 Fair.

Onondaga Historical Association

Fruit monument at the 1917 State Fair.

Coming off the second turn, approaching the half-mile pole.
Canal Museum of Syracuse, William Jubb.

Army tents of the "Fighting Ninth" in 1917.
Photo by Harlow D. Curtis

Civilian soldiers learn their new craft in the bayonet pit.

Onondaga Historical Association

A formation of sad sacks are critically surveyed by a drill sergeant in World War I State Fair encampment.

Onondaga Historical Association

Army stores at State Fair camp in 1917. Note Onondaga Lake in background.

Photo by Harlow D. Curtis

Enterprising merchant offers to take troops to the cleaners in 1917.

Photo by Harlow D. Curtis

Interior of the Liberal Arts Bulding.

New York State Fair

CHAPTER IV

Flamboyant Years

Almost overnight the wartime economy reverted to peace. At the fairgrounds, tentage and temporary buildings, the bunks and other recoverable equipment were soon salvaged, and the bare bones of the former grounds were once again exposed. Since the fair had continued during the war years and all structures were under military maintenance, neglect to buildings had been minimal.

But only the outer similarities could be recaptured; the vegetable displays, the horse and cattle shows. It was an age of recklessness. The war-borne psychology of a nation which was to bring on women's suffrage, prohibition, the "Roaring Twenties," and eventually the Great Depression was, in subtle ways, to change the fair for all time. The demobilized soldiers faced an economy in transition and a shortage of jobs which wasn't overcome until factories could gear up to produce the consumer goods so much in demand.

Emily Estey, a reporter for *The Post Standard*, wrote of the fairs of this period in her book, *Papa Was Positive*, "We never missed a fair. And I can remember all the way back when three dollars was big fair money!"

When the family arrived at the grounds Emily could recall the barkers shouting, "'hurry-hurry-hurry,' and I did my level best. But, a new hitch on a potato digger, now there was something the men thought was really worthwhile . . . when what I really wanted to see were the ponies parading with their little sulkies."

The fair, in Emily's eyes, was "largely a pretend. You look at other people's canned goods and fruit displays and you say, 'We had some as nice as that!' You see the stock in the big barns and you say, "Ol Bess oughta' be here!'

"And on the midway you pretend you are little again and you throw balls at a tin you are pretty sure you won't hit because it makes you remember when you believed you might win the big prizes."[1]

In a way, this make-believe reflected the way a war-weary generation pretended the time was 1912 — or 1913 — or 1914!

1919

The first day of the Peacetime Fair attracted a record turnout of 44,064, shattering all previous marks.

Charles S. Wilson, the State Commissioner of Agriculture, had something up his sleeve. With a new horse barn now a permanent part of the grounds, he announced plans for a new $450,000 Agricultural Hall directly opposite the main gate and at the far end of Empire Court. It was an idea however, who's time had not yet come. Many years would pass before the structure went up.

The Ka-Noo-No Karnival highlighted by a parade of cars, was back. And so was automobile racing.

Just in case someone remembered the circumstances of 1911, the promoters ran an ad: "Let it be said that the precautions for the safety of spectators at these races will be along the scientific lines suggested by the experience of veteran managers of such contests. The motor racing at the Fair will be under the personal direction of J. Alex Sloan of the International Motor Contest Association who will supervise every field and track detail. The day of accidents at these professionally managed competitions of motor fliers has long since passed."[2]

Six days of Grand Circuit harness racing with purses totaling close to $50,000 were arranged, with 30,000 persons watching the trotters on the third day of the fair as two world's records were set. Opening day, however, proved more dangerous than did the auto races at the end of the week. Edward F. "Pop" Gears and Pat Cherrier both suffered serious injuries in the Grand Circuit events.

On Saturday the auto race program once again took its accustomed spot as the closing day feature. The track crowd was announced at 50,000, rather difficult to square with 42,685 admissions officially reported for the entire fair that day.

Louis Disbrow took two of the three heats in the 30-mile main event to be declared winner. As he came to the finish line his car was surrounded by hundreds of his fans who literally tore everything detachable from his car, and spirited away the pieces.

Wartime technology had brought on the air age. No longer were airplanes flimsy, open-framed wonders, but rugged craft with heavy duty engines designed for front line service. Such were the Curtiss Jennys, tough two-seater planes which were readily available to service-trained pilots as Uncle Sam got rid of his surplus, machine and human.

Consequently, a new kind of daredevil emerged, the barnstormer who flew from fair to fair to put on aerial exhibitions and to take up passengers for a couple of dollars.

Such was the two-plane "flying circus" accompanying Lt. Omer L. Locklear, an aerial acrobat whose stock in trade was to change planes a thousand feet in the air by hopping from the upper wing of one craft to a rope ladder dangling from the plane above. All this was accomplished without benefit of parachute.

The entire fair proved a success, with 215,454 visitors, an all time record turnout.

THE SAGA OF TEX McLAUGHLIN

The era of the daredevil was just beginning. New heights of human daring (or stupidity) were to be achieved in the years ahead, with the Fair's visitors to witness much of it.

The air show in 1920 was performed by Tex McLaughlin, who took over the act of Lieutenant Locklear who had been killed in the interim. Tex was fearless, hanging by his teeth, by one hand or by his knees or toes from a ladder between the planes. The real hair raiser however, was when he dropped from one plane to the other, landing on the wing.

Tex's mother, Minnie, who hadn't seen her son for 10 years, came to Syracuse for a reunion and to watch him work.

He was to have a warm-up act.

The final day's auto races, substantially expanded from the sketchy menu provided in 1919, included a series of motorcycle events followed by 10-, 20- and 50-mile auto runs featuring the incomparable Ralph DePalma, along with Indianapolis winner Gaston Chevrolet, Jimmy Murphy, Bennett Hill and Eddie O'Donnell.

Three world's records were broken as DePalma, driving his Ballot car, lapped the entire field to win the 50 miler in 40:49.63, well under the dirt track record of Bob Burman set in Bakersfield, Calif. in 1915. DePalma took Chevrolet by 200 yards in winning the 20 mile event in 16:09. His third mark came in the 10 mile run when he smashed Barney Oldfield's record with a 7:45 performance.

But for once the racers were outshone by the aerial acrobats.

McLaughlin and his flying circus arrived during the last race, just as Eddie O'Donnell escaped injury when he hit the fence, taking out four lengths of board. Tex laughed, saying "Let 'em have that job. I'd rather have mine. It's less risky."

A gusty wind was making steady flight difficult by the time the races were over, and the flyers were to put on their show.

The planes took off, maneuvering into position at 1,000 feet. The second craft loomed over Tex's plane while he waited on the upper wing.

The ladder swung close to his arms several times. Then, on the fourth try, Tex reached up and caught the swaying rung.

Suddenly his plane, being flown by Lt. Shirley Short, one of the pilots for the late Lieutenant Locklear, caught an updraft. The eight-foot-long wooden propeller was turning at 1,400 revolutions per mintue.

The spinning blade caught Tex in the back.

It slashed deep, then disintegrated.

Tex refused to quit. Gamely, he hung on his ladder, then continued to climb to the top rung. The pilot, Jimmy Curran, cut the engine and aimed his plane toward a small field two miles off.

Lieutenant Short's plane, with its prop gone, dropped rapidly, heading for the same field.

Tex couldn't get a grip on the lower wing, and as Curran brought the plane over a row of trees and onto the pasture, the acrobat hit the ground first. He was dragged for several yards before the pilot risked his life in turning the plane sharply onto its wing, thus preventing more damage to the injured man.

Short's plane stopped within feet of the other craft.

Tex could still joke as he was rushed to Crouse-Irving Hospital where 50 stitches were taken to close the wound. He rallied, spoke with his mother, and it was hoped he was out of danger. The injuries to the flier were more severe than originally thought, however. Two days later Harry J. "Tex" McLaughlin was dead.[3, 4]

The 1920 fair had ended on the tragic note becoming all too familiar to fair officials.

Even the trotting races were again haunted by injuries. Ed Geers, who had been hurt in the previous year's racing, was once again sent to the hospital after a collision. He now held the questionable distinction of having "been in more mishaps than any other driver now racing."

THE COLISEUM

One of the greatest advances to the fair (and to the Central New York sporting community) occurred in 1923 when the legislature approved a bill calling for half a million dollars to build an arena at the grounds, even though a similar measure had previously been defeated.

Contemporary reports credit William H. Kelley, the Democratic leader of Onondaga

County, with a major role in aiding the bill's passage which quickly gained Governor Alfred Smith's signature.[5]

It was none too soon. The arena, quickly named the "Coliseum," was a necessary factor before the World's Dairy Congress could be sold on coming to Syracuse.

So, on October 5, and for the next seven days, Syracuse had what might almost be called a second state fair that year, only bigger.

Oscar F. Soule, head of the Syracuse committee, was in charge of the Dairy Congress operation which attracted cattle and dairymen from 40 countries. Blue ribbon winners from every state fair in the nation were present, with hundreds of head of cattle filling every available fairgrounds building.

There were more than simply cows. Band concerts, horse shows and events for boys and girls assured a major turnout.

Henry Wallace, then-secretary of agriculture under the Coolidge Administration (and later vice president under Franklin Roosevelt), joined other leaders on a special train from Washington, via Philadelphia where other aspects of the Congress were held, and thence to Syracuse for the climax.

Under floodlights and "myriads of incandescents," the State Fair Coliseum was opened before the world by Governor Smith.

Five special trains with guests came from throughout the country. A trainload of 1,000 dairy enthusiasts rode up from Philadelphia. Cortland County was represented by 5,000 persons who came in a fleet of 1,000 automobiles.

But the stars were of the four-footed variety. The finest cow in the world, Darling's Jolly Lassie (owned by Ovid Pichard of Marion, Ore., called the greatest living dirt farmer, breeder of dairy cattle), was prominently displayed. A four-year-old Jersey all-breeds champion, she had produced 1,141.28 pounds of butterfat in a year-long test, equal to more than her own weight.

Besides the champion, there were many unique exhibits. Governor Smith was modeled in butter. A giant milk bottle washing machine which could handle 3,500 glass containers an hour, chugged away alongside entire plants for pasteurizing milk.

The spick and span Coliseum was standing-room-only with 5,000 persons waiting for the greatest event of all . . . the million dollar mile of cattle, paraded proudly by owners and Boy Scouts.

The caravan was led by a Holstein delegation, scrubbed and polished until their coats shone like silk. "Bovine lords and ladies of every ilk and color and breed," was one description of the scene on the tanbark floor.

Some of the more exuberant, or impetuous of the cattle, delighted the crowd with their antics, which somewhat disturbed several of the uniformed scouts from the city who really didn't know how to handle the strange beasts they'd been charged with leading.

One city dweller wasn't perturbed. When things quieted down, the governor, a product of New York City's East Side, left his flower-and-flag-bedecked rostrum to join the parade.[6,7]

ANOTHER YEAR OF TRAGEDY

The 1924 fair started off on a high note with dozens of sorority girls selling tickets, vaudeville shows featured in the evening, and polo added to the infield attractions, It was to end, however, on melancholy notes of tragedy and irony, all associated with the track.

For the first time in Central New York's history, it was claimed, polo was played on the specially graded and sodded infield, with the green-clad Madison Barracks Officers' team scoring a convincing 3-1 handicap win over the orange and blue of the Genesee Riding and Polo Club. (Four other soldier goals were erased via the handicap). The officers' horsemanship was described as inspiring, with their mounts in particularly good condition.

Exhibitions were geared to a broad range of interests: the New York Central brought in one of its H-10 steam locomotives; the Post Office Department set up a complete substation, displaying every stamp ever issued by the United States; the midway featured the Patagonian Giant, born with two heads; Jolly Dolly who weighed 634 pounds and Maza and Haza, the joined-together girls.

Even though the weather opening day was threatening, the city was described as "a deserted village," with 48,138 visitors at the fair.

There were some laughs. A state trooper fractured his jaw while eating. A wise guy remarked, "I thought there wasn't supposed to be any hard stuff sold."

The pyrotechnic displays of earlier fairs, including the dramatic "Pompeii" show of 1912, was dwarfed by the fireworks and electricity generated by the exhibition depicting the great Tokyo earthquake of 1923. Forty-eight million candlepower of lighting was said to have been generated in the glistening display which rose 145 feet in front of the grandstand.

But some disquieting news had been forthcoming from other areas, one even before the fair started. On September 1, Race Driver Joe Boyer of Detroit was killed in a 125-mile-an-hour crash on the board track at Altoona, Pa., while trying to overtake Jimmy Murphy. Murphy, the Indianapolis winner and leading for the national championship, ended up the victor.[8]

Then word came from the Wheeling, W. Va., fairgrounds that "Pop" Geers, then 73 and still driving, was killed when he was thrown from his sulky as his horse stumbled. He'd been hurt in Syracuse races in 1919 and 1920.

Like Boyer, Geers would have been a likely starter at Syracuse.

The fair's week of Grand Circuit racing had attracted 200 horses. At stake was $65,000 in purses.

Another Murphy, Thomas G. Murphy, the pride of Syracuse and one of America's premier harness racers, was present and looked toward a busy and remunerative week. It worked out somewhat like that for him.

He and Thomas G. Hinds of Goshen were engaged in a torrid battle in the second heat of the 2:05 trot event. Murphy, with Clyde the Great in the traces, was followed closely by the Goshen man driving Bonnie Del.

What happened then was never clear. Suddenly the two rigs were in a flying jumble of men, sulkies and horseflesh. As his cart flew into the air, Hinds was thrown backwards onto the track. The hooves of one of the horses just cleared Murphy's head. As bystanders shifted the wreckage, Hinds was found dead of a broken neck.

Murphy went back to the stables and with a new sulky drove Hope Frisco to a second in the 2:10 trot, then reharnessed Clyde the Great which he guided to victory in the third and final 2:05 heat.

A little while later James Lewis, 51, of Syracuse was passing the Murphy stable when a bale of hay fell, crushing him. He was taken to the Hospital of the Good Shepherd in critical shape.

Through it all, Tom Murphy never lost his composure even though it was later disclosed he'd broken a rib in the collision. His fair racing wasn't over yet. Murphy closed out his stay with a win in the 2:08 trot for $10,000 with Tillie Brooke defeating Pluto Watts by a length, and succeeded in winning three major stake events in a single afternoon![9]

The auto racers arrived in Syracuse in mid-week to prepare for the 150-mile national championship. Rain forced postponement to Monday, September 15.

Starter Fred Wagner flagged off the field, including many of the world's finest: Tommy Milton, Jimmy Murphy in a bright gold Miller Special; Phil Shafer in a Dusenberg Special; Benny Hill, also in a Miller; Harry Hentz in a Durant Special and Earl Cooper in a Studebaker Special.

Murphy was in the third cluster of cars to get away and worked his way to second by the 15th lap. He dropped behind to third in the next circuit, then fell further to the rear. He recovered third on the 35th lap and succeeded in once again taking over second spot on the 68th tour when Tommy Milton dropped out with tire trouble. Even so, Murphy had been lapped three times by Shafer who was having an easy ride.

On the 138th lap, Murphy's car reached the vicinity of where the Hinds trotting wreck had occurred. The car lost traction, then lurched into the infield railing. Seventy-five feet of the board was torn away. A section rammed through the car's hood. Another plank hit Murphy.

Benny Hill swerved to the outside, just missing the golden Number Two. Harry Hentz came on the scene and also skidded into the fencing, but with the help of spectators got back on the track and resumed the race.

The run continued. Shafer won handily in 1 hour 54:25.20, a world's record.

It was at 4:45 p.m. that Jimmy Murphy, the national champion, was pronounced dead at St. Joseph's Hospital. He was the first race driver to die on the fairgrounds track.[10, 11]

THE HAMBLETONIAN

The spectacular times recorded in the trotting and pacing races, the support of local fans and the superb barn and transportation conditions resulted in Syracuse being selected for the running of the first Hambletonian Stake in 1926. A purse of $73,600 was established for what was planned as the sport's premier event, the Kentucky Derby of trotting.

The race was open to three-year-olds, and with the victory worth $45,868.42 along with the prestige for stud fees, the finest of the nation's harness horses were entered. The final listing included 14 trotters from an original list of 600. It was termed "the most prestigious event ever attempted in the history of light harness racing."[12]

Race day was perfect, with a brilliant sun, blue sky and cool breeze. The grandstand, bleachers and rails were packed along the front straight and railbirds crowded halfway through the turn, with more than 34,000 enthusiasts in attendance.

The races were half an hour late in starting, but at 2:30 p.m. the crowd was rewarded by seeing the contestants trot through several warm-up sprints, then line up for the parade before the stands.

The greatest ovation went to Syracuse's Tom Murphy wearing red, white and blue racing colors behind Full Worthy. He had warmed up by winning a three-year-old pace heat with Highland Scott.

Guy McKinney, a long-legged bay colt, son of Guy Axworthy and owned by Henry B. Rea of Pittsburgh, was established as an 8 to 5 favorite. Nat Ray of Cleveland, O., was the driver. At the word "Go," McKinney swept in front with Guy Dean, another Pittsburgh colt and Axworthy descendant on the pole, maintaining a close second. As the field swung into the backstretch, it was McKinney in full command.

Guy Dean pulled up even at the three-quarter pole, but with a furlong to go McKinney called on his superior power and went smoothly to a 2:05¼ half-length victory.

In the second heat McKinney bowled along with its head flung straight out, long black mane flying, to again win over Guy Dean by a half-length. The winner's time was 2:14¾, well off the track record of 2:03½.

Full Worthy finished fifth in both heats.[13]

For several years Syracuse was the home of the Hambletonian.

As *The Post-Standard* declared in an editorial, "Among harness races the Hambletonian holds a peculiar and conspicuous place. And the Hambletonian draws people." Syracuse was proud of its big race.[14]

Then it was gone, shifted to Good Time Park in Goshen.

AN INDIAN VILLAGE

An important addition to the fair came about in 1928 when the Indian Village was dedicated. It was located at the northwestern corner of the grounds in a pleasant stand of elms. Several bark huts were erected at the center of the clearing, with the entire village planned by Dr. Erl A. Bates, ethnologist at Cornell University, with the cooperation of the Six Nations of the Iroquois Confederacy.

The "Corn Marathon," a run from the Onondaga Reservation south of Syracuse over several steep hills to the fair, was conducted to mark the village's opening. Members of the various tribes participated in the race, with Chief Andrew Gibson at the Village's Bear Mound to greet the runners as they crossed the finish line.

Andrew Lewis, the winner, was timed in 2 hours 15 minutes. Others among the 14 participants were Andrew Williams of the Cayugas, Percy Smoke of the Senecas and Jefferson Thomas of the Oneidas.

The Indians had to share their activities with the aerial heroes and heroines. The arrival of Aviatrix Amelia Earhart, the "Lady Lindy," brought thousands out to the grounds. Others waited in vain for the arrival of the dirigible "Los Angeles," which failed to make its scheduled fly-over.

The disappointment was forgotten by 1929 as America's fascination with flying machines reached a new high. The daring flyers of the National Air Races, the Powder Puff Derby girls and the thrilling achievements of the Graf Zeppelin in its aerial cruise around the world under the command of Hugo Eckner, were front page news as the '29 fair opened.

The show's promoters were quick to catch the mood, and the airplane was once again a feature of the exposition.

There was also Wilhelm Weidrich who flew, but without an airplane. He was better known as "Wilno, the human cannon ball."

"Wilno" thrilled the crowd at the grandstand by being shot from a giant cannon located in the infield to a net supported by a steel framework several hundred feet away.

They got more than they bargained for at the evening show on August 28, when "Wilno's" cannon failed to hurl him far enough. The unfortunate daredevil fell just short of the net, his chin striking the frame's iron crosspiece nearest the cannon.

Suffering from head and shoulder injuries he was taken to the hospital where his condition was soon found to be not serious. The next day, after repairs were made to the cannon, the flights were resumed . . . with brother Hans taking over as the projectile.

ENDURANCE FLIGHT

The aerial program included something special — an attempt by Clyde E. Pangborn and Carl A. Dixon to set a new world endurance record for continued flight, the trip to be accomplished over the fairgrounds.

World record endurance stunts, whether they were in aircraft, on bicycles, on the dance floor or atop flagpoles were crazes of the times, as exciting to bystanders and participants alike as Guiness Book of Records attempts are today.

As fairtime approached, local readers learned of the fate of one record holder, Thomas Reid of California, who died when he dozed off after setting a new world's mark for solo endurance and flew his plane into a tree. He was seeking a promised $100 for each additional hour he stayed aloft. Reid wanted the extra money to pay for his honeymoon.

An airline pilot complained about the stunt flying of Colonel Charles A. Lindbergh. And Lieutenant Jimmy Doolittle's life was at stake as the wings of his plane folded during a dive. He parachuted to safety from 2,000 feet.

Meanwhile, Clyde Pangborn arrived in Syracuse. He was a veteran of 11,000 hours in the air since he began barnstorming in 1919, had been a leader with the Gates Flying Circus and had carried more than 100,000 passengers.

A crowd of 5,000 gathered at the old Amboy Airport west of the fairgrounds (now part of the Solvay Process wastebeds) to watch Pangborn and his Empire State Standard single-engine biplane with its bright orange wings and green fuselage get in some aerial refueling practice from airmail pilot Cy "Shorty" Bittner's plane.

The craft was especially designed for the flight, with a cot under the cowling in the front cockpit. Both cockpits were open, the pilots wearing woolen clothing to keep warm. Two separate oil tanks were installed so oil could be changed while the plane was in flight. A catwalk was located entirely around the motor in order that the pilots could creep within inches of the whirring propeller to make repairs.

Refueling plans were developed as follows:

As Bittner flew above, a long rubber tube would be dangled where Dixon, in the front cockpit of the Standard, could grab it and feed it into a pipe leading to the gas tank, a tricky maneuver with the ever-present danger of gasoline flowing onto the hot engine.

The ship was christened by J. Dan Ackerman of the state fair in front of the Empire Air Transport Hanger. As he smashed a bottle of champagne over the prop's hub, the prohibition-wise crowd was quick to note by its odor that the liquor was "the real McCoy."

It was announced the flight's goal was to beat the record of 420 hours set by Dale Johnson and Forest O'Brine in the St. Louis Robin, a cabin plane. Takeoff was for 9 a.m. Monday, August 26 at the race track infield, with the flight to conclude on Friday, September 13. All refueling contacts were to be made in full view of the stands.

Meanwhile, the Graf Zeppelin was on its victory tour following the round-the-world flight and was scheduled to fly over Syracuse in a few days.

At the appointed time the Empire State Standard was loaded with 90 gallons of fuel in tanks built to accomodate 200 gallons. It was decided to keep the takeoff load to a minimum because of the short infield runway.

The engine was revved, then the plane rolled across the grass strip. In 15 seconds the wheels were off the ground. Pangborn set a southwest course over the 25,000 persons watching the takeoff.

The flight soon settled into a routine, day and night, sunshine and rain. By 2:33 a.m. on the second day, the Empire State was purring smoothly through murk and light rain. Like a gigantic black-winged night bird it wheeled and circled around the airport, zoomed over the eastern part of the city, then turned west again, sailing monotonously on and on.

The pilots alternated at the controls. On the first day they made four contacts with the Uneeda. Supplies, including shoes for Pangborn, soap and towels and a pocket checkers set were lowered. The food, served via rope at 5:21 p.m., included vegetable soup, roast beef, mashed potatoes, peas, ice cream, cake, fruit, milk, coffee and water. Immediately afterward the pilots passed along a message:

"Don't put liquid foods in cardboard. They collapse and leak."

Robert R. Mill, a reporter, went aloft for a special visit, with his pilot jockeying to a spot less than 10 feet from the Standard's wingtips. The endurance team pointed proudly to their clean-shaven faces.

Aaron Kranz, the "Diavalo," was a drop-in visitor, climbing down from the Uneeda to the Empire State then, after a few minutes' chat, dropping away in his parachute.

Before he left Kranz, an experienced mechanic, tried to fix a cracked exhaust manifold which had caused fumes to blow into the pilots' faces.

During the long nights and short naps, the flyers experienced odd dreams. Pangborn found himself in a strange town where he met Charles Lindbergh's brother (unusual, since Lindbergh didn't have one). The town had a house everybody dreaded. Undaunted, he went in to discover an old woman with a vacuum cleaner . . . into which all strangers were drawn.

Dixon's dreams were more relevant. He'd wake up thinking they were in a spin.

Not that there was too much sleeping. The bunk in the front cockpit wasn't satisfactory and the tradeoff for an open cockpit to save the additional weight of a cabin meant a considerable sacrifice in comfort.

Thousands of area residents kept them company one night however, as they waited up to catch sight of the Graf Zeppelin enroute from Chicago to Lakehurst, N.J.. But, like the Los Angeles the year before, it was all for nothing. Troubled by forecasts of thundershowers, the dirigible sailed well south of Syracuse.

The Standard droned on until the evening of September 7. Dixon was at the controls, circling the plane while Pangborn caught a few winks. Glancing at the gauges, he noticed the oil pressure dropping. There was a smell of smoke.

He roused Pangborn, then climbed out onto the catwalk. The problem was easy enough to spot. Oil was spurting from a connection where a small piece of rubber tubing had split. Dixon tried to insert a new hose, but was immediately blinded by oil. It soon covered the fuselage and Pangborn's goggles.

Stripped to his undershirt and trousers and bathed in oil, Dixon clung to his slippery perch while trying to reconnect the tube. The engine would die down, then pick up.

Two hours passed in the cold night air before Dixon completed the work. Just as the job was finished, the oil tank drained. Before the spare tank could be switched in, the engine seized.

Meanwhile, the plane was a 700 feet over the Amboy field. At that point the navigation light on the lower wing quit. Pangborn signalled to the airport with a flashlight.

While the plane was 400 feet up, the engine cut out for good. Empire Transport personnel dashed across the field and had the airport lights turned on. But by then the craft was almost in the ground.

Somehow the pilot kept the plane upright as it touched down. It was using up field at a rapid rate. Pangborn vainly sought the foot brakes but his legs, cramped from disuse, couldn't work the pedals and the Empire State Standard crashed into a Curtiss Fledgling parked at the field. This innocent craft in turn rammed into the municipal hangar.

The time was 10:45 p.m. The Empire State Standard had been aloft for 179 hours 44 minutes and had covered an estimated 8,740 miles. Neither pilot suffered more than bumps.

Dixon's comment as he jumped from the plane: "Gee, that's a tough break."[15, 16]

The endurance bug spread to others as well. In addition to the exploits of aerial endurance flyers, a golfer, Jerry Damiano, had broken a 24-hour golf marathon record by playing 225 holes in a day. And two boys in Memphis, Tenn., had ridden a bicycle 107 hours. Several Syracuse youths figured they could do even better.

Gerald McDonnell, 13, of 404 Bellevue Avenue and Edward Williams, 14, of 218 Rich Street, pedaled their bike on streets near their homes for days until, with the new record well within their grasp, they headed for the fairgrounds on the final day of the Show. There they received a hero's welcome as they reached a new world's record of 175 hours and five minutes.

But records are made to be broken. Henry E. "Red" Ball and his partner, Harry Francis, bicycled even longer, completing their ordeal of 190 hours 35 minutes at 12:35 p.m. Wednesday, September 4 in front of *The Post-Standard* building on South Warren Street.

They began their journey at 2 p.m. Monday, August 26 in front of 604 Almond Street. They'd had their problems when a rear tire blew on Sunday night. The twosome cut it off and rode on a tireless rim for almost a full day. For the entire period, one or another of the boys was aboard the bike.

What type of bicycle? "Mongrel, I guess," one answered.[17]

Officialdom of course, got into the act.

The bike endurance contest was condemned by Dr. George C. Ruhland, the City Health Commissioner. But despite that, the city was swamped with requests to use streets for the lengthy bike rides. In yesterday's version of Catch 22, Police Chief Cadin sent the kids with their requests along to the Department of Health for final approval.

At any rate, the year of endurance record tries was over insofar as the fair was concerned, even though a short-lived attempt by Aaron Kranz flying out of Amboy ended in a forced landing several weeks later.

BACK TO THE MILE

Despite the flat turns, the mile oval was well-established as the fastest dirt track in the world. The only recognized race record for a mile dirt course not made in Syracuse during the so-called golden years of the 20's was Ralph Mulford's 200-mile mark set in 1919 in Columbus, O., probably because the distance was never run in Syracuse. In all, 22 records were set at the fair during the decade from 1920 to 1929.[18]

It was the last race of the 20's which turned out to be perhaps the most dramatic in the fair's history.

Shorty Cantlon began things with a new one-mile record of 41.2 seconds in qualifying.

Deacon Litz took the lead at the start of the 100, setting world marks at five, 10 and 25 miles, but the pace ended in a blown engine for him. Frank Brusco took over, establishing a new 50-lap figure.

The dust and continued passing and repassing soon threw officials and spectators into confusion. As the race moved into the last six laps, fans were cheering relatively unknown Fred Winnai in a fire-engine red car as it refused to budge for Wilbur Shaw who was camped on his tail. As the checkered flag fell, it was Winnai who was escorted to the Winner's Circle.

A recheck of the scoring some four hours after the race was over revealed Shaw, instead of being a scant hood-length behind Winnai, was really just seven feet short of a mile ahead.[19]

Shaw was to go on to win three Indianapolis 500's in the next 10 years.

THE 1930'S

Conduct of the fair in the decade of the 1930's was especially affected by outside circumstances, among them the great depression affecting all Americans; the politics of state government and the influence of the landmark Chicago World's Fair of 1933.

After the Daniel Parrish Wittier Agricultural Museum was constructed at a cost of $50,000 in 1928, the next structure to be erected was the Boys' and Girls' Building in 1930. Located on the northwestern quadrant of the grounds near the Indian Village, it cost $150,000.

The nation's economic problems resulted in a further dropoff of income and legislative appropriations for capital construction. Some help was forthcoming from the state's Assembly and Senate. Expenses for the fair could now be paid directly from gate receipts and entry fees, the balance to be used for new buildings rather than the old custom of turning everything over to the state controller and the general fund.[20]

The fair lived up to an earlier promise to the women of the state when a new Women's Building (now known as the Center of Living Building), an attractive colonial-style structure along the eastern side of Empire Court, was dedicated on September 6, 1934. Doing the honors was Mrs. Herbert H. Lehman, wife of the then-governor. It wasn't too many years earlier that women's events were housed in tents and, on one occasion, in space which canine exhibitors had just vacated.

In the late 1920's a committee headed by Ann Phillips Duncan, Vera McCrea, Harriet May Mills and Eliza Keats Young sought sanction for a more suitable building to handle the growing numbers of programs devoted to women's interests. They found a supporter in Charles H. Baldwin, then-Commissioner of Agriculture, who urged the Agricultural Council to back the project. Governor Franklin Roosevelt gave his o.k. and eventually money was forthcoming. With Roosevelt going to the White House, the final approvals were granted by Governor Lehman.

The money was also appropriated for a new $239,000 Agricultural Building, a massive Farm Machinery Building for $175,000 and a Pure Foods Building costing $112,000.[21]

The management next turned to needs for landscaping the area in keeping with the immaculate appearance of the Chicago Fair; the streamlined architectural renderings of the upcoming New York World's Fair and to overcome the muddy realism of a Syracuse Fairgrounds saturated with new construction. Under the direction of Paul Smith, a major effort was made to spruce up the entire facility for the 1938 and 1939 shows, with emphasis given to better exhibitions, headline attractions and floodlighting for nighttime programs.

IT BECOMES AN EXPOSITION

As a psychological ploy, the fair was termed by 1938 as having become more . . . much more . . . than simply a "fair," and the formal name was changed to New York State Agricultural and Industrial Exposition.

The idea wasn't exactly new. It had been verbalized in the early 1900's by then-Governor Hughes who said the state fair could never develop into an exposition until provision had been made for the recognition of factory products on the same scale as products of the soil, and until it recognized the interdependence of industrial and agricultural activities.

The public ignored the platitudes. To them it was still "the fair."[22]

Traffic also continued to be a trial to fair administrators. A county resident described, as though it was for the first time, a typical state fair traffic jam of the mid-30's.

"Imagine," he said, "the State Fair Boulevard, a distance of about two miles — packed four deep with cars, moving a few feet every thirty or more seconds; the air filled with gasoline smoke, to say nothing of the terrific stench of the Tarvia plant and the 'sunken gardens' of muck and filth along the road.

"Such was the picture that presented itself to families and others who drove to the fair around noon Monday . . there were crying children, fretted mothers, cross and irritable fathers. There were 'fresh' drivers who took delight in rendering the situation more unbearable by the honking of horns. There was the man ahead who insisted on racing his motor, filling his neighbor's car with smoke and fumes. There was the car which had run out of gas and had to be pushed along by those behind. All in all it was terrible."[23]

PLANNING THE MONSTER FAIR

Great and heady dreams reached the paper stage in 1937 as a committee of 100 men and women was established under the direction of Joseph A. Griffin to plan for a monster 90-day show to celebrate the 100th birthday of the state fair in 1940.

Among the physical changes suggested was a massive landscaping of the entire area and enlargement of the facilities to accomodate a major airport with four intersecting runways along the stretch between the fair and Onondaga Lake; a bathing beach, harbor and hydroplane base along the water's edge; a tourist camping area; a permanent midway; traffic circles, terraces, flower beds and lawn areas. The race track would include lagoons in the infield.

Griffin said the fair's Centennial "should be commemorated with a mammoth fair and expostition that will last for three months . . . Down in New York they are planning a world's fair for 1939 . . . There is no reason why our centennial should not be the largest exposition of its kind ever staged in Upstate New York."[24]

The decision by the World's Fair Authority to continue its New York City operation for another year, plus a lack of money and support for the tremendous capital outlays needed, resulted in the state fair blueprint being relegated to the back of the file drawer. It's still there!

THE LATE 30'S

More realistic efforts continued to improve the fair during the last years of the decade. It was the era of big bands and jitterbug dancing competition. The strains of Guy Lombardo and his Royal Canadians and the Benny Goodman Orchestra attracted thousands to the two-week fair of 1938, and continued on during 1939 as well.

One part of Griffin's great blueprint did come true, if only for a couple of years. Introduced was a championship steeplechase course in a garden setting within the infield, so that a full slate of running races could be held in addition to the harness events. Two lagoons complete with swans and ducks (but no flamingoes) graced this area.

An aviation show limited to on-ground displays of a million dollars in parked planes and equipment; auto and motorcycle races, and opportunities to hear "The Messiah," see a rodeo and thrill to the exploits of Lucky Teter vied with Homer Rodeheaver, the community song leader, in a 2,000-voice singing program.[25]

The popular "Big Cheese," weighing six tons, returned to the Fair in 1937. "Built" by W. Clarence Kelsey of Copenhagen, its statistics are still impressive:

Needed were 61,000 quarts of milk (at a cost of close to $30,000 at today's retail prices).
Height was seven feet.
Diameter was six feet.
Circumference was 19 feet.
Time required to make: one day.
Time required to cure: seven weeks.
Type: Cheddar.
Equal in energy to 81,000 pounds of chicken; 30,000 pounds of beefsteak; 30,000 pounds of eggs or 96,000 pounds of peas.

Old timers remembered however, that in 1921 a 22,000 pound cheese monster had been produced by Kelsey. A special truck with extra framing and springs had to be built to carry the monumental curd.

In a nearby building the State College of Forestry unveiled its latest creation. The institution which had turned out some 500 displays since a first showing of creosote fence posts in 1913 had by 1936 hosted more than six million visitors at its exhibits.

But its greatest achievement came during the 1939 fair when the college and the State Conservation Department jointly created one gigantic exhibit: "How to Attract Wildlife," in a variety of situations.

Professor Ralph King and his colleagues in the Department of Forest Zoology and Roosevelt Wildlife Station worked up dioramas representing an opening in the forest; a pastured woodlot; a fence row around a cultivated field; roadside and idle land; a managed woodlot; cultivated fields in the fall; a home orchard in fall; a summer camp; a golf course; a lawn and garden; open water; a lake bottom and shore; a stream and a spruce bog. Dr. Justus Mueller of the faculty spent the entire summer painting 1,400 square feet of diorama backgrounds. Huge trees were brought in from the Huntington Forest in the Adirondacks by the Public Works Administration and college workers, and tons of earth were trucked in. One hundred and forty eight thousand people viewed this one display![26]

During the decade the race track faced up to a new generation of racing cars, with riding mechanics accompanying the driver in competition, thus doubling the risk of injury when an accident occurred. It was an era when safety was of secondary consideration, especially for the car crews. Roll bars or cages, seat harnesses and fire retardant suits were non-existant. Even helmets were rudimentary. Without fuel bladders, a split gas tank was an invitation to fire. Engine performance had far surpassed the rest of the state of the art.

On September 12, 1931, Jimmy Gleason of Philadelphia, a top racer, was killed when his

Duesenberg Special rammed the concrete wall at the end of the front stretch near the first turn, while his riding mechanic, Earl Younger of Indianapolis, was seriously hurt. Seven spectators were also injured.[27]

Several years later, in 1934, the racing commenced on a light vein, with the Ancient Benzine Buggies Show's two-mile feature race taken by Ray Smith in a 1908 Hupmobile at a bone-rattling speed of 31.86 miles an hour. Of the five cars entered, only three — a 1906 Ford, a 1907 two-cylinder Maxwell which developed a burst of speed carrying it to 20 mph under Driver Richard Wright, and the Hupmobile finished.

The 100-mile A.A.A. race was rained out and carried over until Sunday. It was to cost the life of George W. Brayen, a 30-year-old Barneveld, N.Y. garage owner and former Cazenovia resident, and serious injuries to Mechanic David Damon of East Syracuse when they smashed into the fourth turn wall.[28]

In time trials another wreck in almost the same spot caused serious injury to Bill Gorman, a riding mechanic, and a confrontation between the police and a Syracuse newspaper photographer which today would have had tremendous implications.

Herman Borzner of *The Post-Standard* was on the scene within seconds, taking pictures of the smashed car driven by Ed Kessler of Buffalo. He was told to leave by a policeman, but continued to use his Speed Graphic. The trooper, armed with a lead-filled baton, rapped Borzner on the wrist, then hit his camera.

A small riot broke out as the photographer's camera was seized and for a few moments the situation threatened to become a full-blown constitutional battle over the rights of the press at a public event.

Within a few hours the incident was resolved and the orders preventing photos from being taken at wreck scenes finally countermanded. (It is interesting to note that Borzner's son eventually became a state trooper).[29]

Once again tragedy hit in the 1940 race, this time in the 17th lap when Lou Webb of Los Angeles, driving a Marks Special, ran over the wheel of Kelly Petillo's car in the east turn. Webb was thrown clear as the car spun high in the air and broke up on hitting the track, but he died in the ambulance.

Rex Mays of Glendale, Calif., a perennial Syracuse favorite, won the race.

A LAMENT FOR WORTHY SOUVENIRS

A lamentation was sounded during the 30's on the lack of worthwhile souvenirs at contemporary fairs.

The editorialist mourned that "No longer do entranced children stagger home at the end of a long and joyous day, burdened with the weight of countless gifts.

"True, the inevitable yardstick was present . . . but where were all the gadgets of yesteryear?

"To the children of a former generation, the fair was not an episode of a day, but an event recreated daily for weeks as the treasure bag full of souvenirs was brought out and its contents fondled much like King Midas fondled his gold.

"In those days the firms showing farm machinery, silos, barns and other agricultural equipment passed out golden fish to every grasping, childish hand. Place them in the palm of the hand and the natural heat of the body caused them to curl and twist in a most entrancing manner.

"Then there were pencils, notebooks, cardboard models from which houses and barns and silos were made, balloons, whistles, models of farm machines and a host of other things spelled TREASURE in capital letters to all youngsters.

"Those days seem to be gone forever, no longer can the children take their souvenirs of the fair to the school yard . . . truly the fair has degenerated . . . there ought to be a law ! . . . what's a single miserable yardstick in comparison to the loss of treasures handed out in days gone by?"[30]

Specially built truck was needed to carry the 1921 Big Cheese which weighed 12 tons.
William J. Delaney Collection

The 1921 racers spring past the starter and infield judges' stand.
Onondaga Historical Association

Motorcycle racing was popular with the railbirds.
Canal Museum of Syracuse

Teamwork at the '24 Fair shows what farming was like in 1824.
New York State Fair

Beauty is in the arms of the beholder as these feathery stars attest.
Canal Museum of Syracuse

The art of the cooper is demonstrated.
New York State Fair

Trotters pass the grandstand before standing room crowd, circa 1925.

The pause that refreshes at the Holstein Milk Bar.

Canal Museum of Syracuse

The log cabin in popular 1929 exhibit.

Canal Museum, Doust

Indian house at Fair in 1929.

Syracuse Newspapers

Then-Governor Franklin D. Roosevelt speaking at 1929 Fair.
The Post-Standard

Wilhelm Weidrich, the "Human Cannonball," enroute to a miss at the 1929 Fair when the cannon's spring mechanism failed to generate sufficient propulsion. His brother, Hans, took over for the injured trouper, who later recovered.
Onondaga Historical Association

View from the Uneedas, showing upper wing and special engine walk on Empire State Standard.
The Post-Standard

Refueling the Empire State Standard from supply plane, the Uneedas.
The Post-Standard

Mission completed. Fuel hose drifts to right as fueling is completed. Note gasoline vapor.
The Post-Standard

*Clyde E. Pangborn,
1929 endurance flight pilot.*
The Post-Standard

*Carl A. Dixon,
co-pilot on endurance flight in 1929.*
The Post-Standard

Wilbur Shaw, helmet, and Mechanic Dean DuChemin shown during winning performance. Shaw won three Indianapolis 500's.
Onondaga Historical Association

Driver George Wingerter and Mechanic Clyde Goeschel, left, were in the field for mid-1920's races.
Syracuse Auto Club

Big cars, with mechanics' seats empty as safety measure, roll toward starting line.
Onondaga Historical Association

Governor Herbert Lehman and his wife at 1934 Fair.
Onondaga Historical Association

The pre-war Fair with old grandstand and trackside packed with auto race spectators. Note trees along State Fair Boulevard. This was prior to establishment of Solvay Process wastebeds along the Onondaga Lake frontage.
Syracuse Newspapers

Main gate in 1940 proclaims Polish Relief Day.
Syracuse Newspapers

State Fairgrounds in the pre-World War II period when lagoons graced the infield, and State Fair Boulevard was a narrow tree-lined thoroughfare. Steeplechase course can be noted in infield, with various jumps delineated.
Greater Syracuse Chamber of Commerce

CHAPTER V

A Time to Rebuild

Although Europe had been at war for two years, the 1941 fair had shown little effect from the conflict. In some ways it paralleled the years prior to America's entry into World War I, when it was business as usual during the 1914-1916 fairs.

AFTER PEARL HARBOR

But the aftermath of Pearl Harbor was to have a far greater impact on the United States, and in turn the New York State Fair. The military once again took over the fairgrounds for troop housing and as a major supply depot. Even so, the grounds never did become the great training center it was in World War I. For one thing, the land simply wasn't there for expansion to house 25,000 soldiers. A large section of the land north of the city had been designated for a major bomber base and the need for desert and jungle warfare training favored southern and southwestern sites, as did the year-round characteristics of those camps.

The all-out aspects of the war economy, early resort to rationing and the tremendous demands of a huge military establishment on agriculture all mitigated against the uninterrupted continuation of the state fair. And, despite military maintenance, the physical plant soon deteriorated. The roofed grandstand was the first to go, a casualty of both war and weather. A storm in June of 1942 seriously damaged the 560-foot-long stand with its seating for 7,500 persons. The finishing touches however, came with a 40-mile-an-hour storm in July. Most of the roof was wrenched off. A metal-starved war economy gratefully accepted the 700 tons of steel which was shipped to the Rome Air Depot.

A new kind of hazard faced the fairgrounds during these years. At 3 a.m. on January 20, 1944 the Solvay waste bed dike at the Powell Farm broke. A wave of whitish sludge rolled across State Fair Boulevard and onto the fairgrounds, flooding 85 acres, the D.L.&W. railroad tracks, Christ Community Church, a garage and several homes.

One of the guards at the fairgrounds, which was then in use as an Air Corps depot for storage of gliders, helicopters and various other materiel, noticed that the station house, borne on an eight-foot-high billow of sludge, was coming right at him!

In a rush to get out of the way, he lost his revolver and his false teeth.[1]

Army jurisdiction continued over the fair until well after the fighting in the Pacific was over, exercising the option it had to continue its lease of the grounds until October, 1946.

But even before that a living ghost of fairs past returned for a visit. It was Ralph DePalma who had won championships on the mile track on three occasions. His once-black hair was silver and instead of coveralls and cloth helmet he wore a conservative business suit as befitted an industrial engineer.

His steps were jaunty and his blue eyes were described "as keen as ever" as he served as official starter for *The Herald-Journal's* soap box derby in July of 1946 before 25,000 Syracusans.²

Racing fans would have to wait several years before cars of the motorized kind returned to the community.

A NEW HOME?

No one in 1946 was quite sure whether there would even be an interest in reestablishing a state fair, especially on its Solvay site nestled between a steel mill, railroad mainline and an industrial wastebed. The Syracuse community had changed dramatically. Its once-modest-sized university now had 16,000 students; new suburban areas were popping up to support a new type of worker — the highly technical personnel employed in locations such as General Electric's Electronics Park, and at the growing medical facilities. Furthermore, there was a question as to what the post-war generation would want in terms of entertainment.

A temporary State Fair Commission recommended that the state spend $52 million to turn the Mattydale Air Base north of Syracuse into a year-round world's fair or exposition grounds with suitable display, recreation and merchandise facilities. The property contemplated was adjacent to the then-proposed New York State Thruway and south of the Hancock Airport east-west runway.

Meanwhile, Governor Thomas E. Dewey, who had been in Syracuse earlier to tour the fair's facilities, turned "thumbs down" on a suggestion that the fair be transferred to a site in Oneida County.³

In a lengthy letter, Walter L. Welch of Syracuse came up with strong support for retention of the present location. He cautioned the thrust and scope of a state fair was considerably different from that of a world's fair, and must be carefully studied before irrevocable decisions were made.

The advocates for a new location listed four basic reasons:

1. Deterioration of the present grounds.
2. Impossibility of expansion.
3. Danger of another waste dike breakthrough.
4. Inadequacy of present roads for fair-generated traffic.

Welch put up four alternate propositions:

1. Would it cost more to rehabilitate the present grounds than to start anew at a different location?
2. If the top paid attendance is about 78,000, "what means are to be employed to insure attendance at the proposed world's fair on a year-round basis to warrant the expenditure of several million dollars of public funds?
3. Why doesn't our representative in the assembly do something about the source of danger from the waste dikes?
4. If the important new highways are to pass by Mattydale, "why wouldn't it be possible to build a connecting road a mile or so further to our present fairgrounds?

He stressed the potential for following the plans of the late Joseph Griffin to beautify the lakeshore wastelands, and their use as a recreational area. He adhered to the philosophy of building a fair through grass roots established "in every town, village and county of the state. Every agricultural society, home bureau, grange, community planning group, research group,

116   Empire Showcase

women's club, health organization, education group, music society, art and social welfare group . . . should be invited to participate in its planning and staging."⁴

With the war three years in the past, the decision to go or not go with the state fair had eventually to be decided in Albany.

THE INTERIM FAIR

The State Fair Commission headed by Clellan Forsythe, a Syracuse Republican and the agricultural commissioner, opted for a limited, or interim fair devoted largely to livestock exhibitions as a way to "break in." This was especially significant since the state's administration was considered "opposed to sinking large sums into a fair site which is scheduled to be abandoned eventually."⁵

The leadership also felt it would be dangerous to conduct many of the traditional events such as racing on the antiquated track.

The six-day limited fair in mid-September, 1948, with Bligh A. Dodds the director, was just that . . . limited in scope and interest. A total of 25,529 persons went through the gates, with a single day's high of 5,431 marked on September 17.

Featured were a rodeo, a horse show sponsored by American Legion Post 41 of Syracuse and the judging of cattle from 38 counties, with almost all events conducted within the Coliseum. This structure had weathered the war in good, though somewhat grimy condition.⁶

Even though the 1948 affair hadn't received rave notices, it did enable the fair officials to assess the possibilities. As they went to Albany in early 1949, the Syracuse delegation to the legislature, including Assemblymen Lawrence M. Rulison, Searles G. Schultz of Skaneateles and Donald H. Mead of Syracuse and Senator John H. Hughes, presented a united front for a full 1949 fair.

At the same time they withheld comment on the $52 million Mattydale proposal.⁷

The moment for a new fair location had come and gone. By January 27, rumblings out of Albany reflected the view of the State Fair Commission that it was leaning towards remaining at the old site.

THE FULL FAIR RETURNS

Planning was thorough, using the sales techniques of past years along with some new gimmicks to assure the best possible turnout.

A motorcade of 50 cars under the auspices of the Syracuse Automobile Dealers Association visited 32 cities throughout the state to extend invitations.

Well in excess of 50,000 advance sale tickets were sold at 35 cents each in area stores and banks.

Blonde, grey-green eyed Phyllis Roderick, a 19-year-old junior at Syracuse University was chosen State Fair Queen over 360 other contestants.

Judy Canova was announced as the celebrity attraction.

And Governor Dewey was assured as guest speaker at both the traditional Farm Dinner (once again reinstituted at the Hotel Syracuse) and at Governor's Day.

Although the '49 fair was again limited to six days, opening on Labor Day then closing the following Saturday, officials were aiming for record-breaking one-day and overall attendances. The one day record had been achieved on Labor Day of 1938 when 78,299 came through the gates, while the total overall record was 380,000 for a nine-day fair.

Hot, sunny weather greeted tens of thousands of New Yorkers starved for a fair after eight years' involuntary abstinence. As they poured into the old Empire Court on Labor Day, they proved the decision to keep the fair at the same site hadn't been all wrong even though the plant and highway network left something to be desired.

The visitors came by cars, trucks and buses, creating a five-mile-long jam extending all the way downtown. Thousands waited patiently in long lines at the bus terminal at West Genesee and Clinton Streets. By 10 a.m. the attendance had already smashed the full week total for the previous year!

Police officers, city, county and state, all agreed it was the largest traffic jam in the area's history. Packed buses were emptied as riders walked the mile or two to the grounds, further compounding the problem. Some 12,000 riders took the buses that day.

Automobile traffic to the fair was finally cut off completely at 4:45 p.m. and wasn't reopened until 7:30 p.m.

The gatekeepers commenced their tally. By the time they'd added up everybody, the official total was listed at 103,650 . . . an alltime record which was to stand for decades.[8, 9]

Unfortunately, many fairgoers never did get there in time to see the featured attractions including the first stock car event ever held on the old oval. The NASCAR-sanctioned races included cars which "must have fenders and running boards, if they were originally so equipped. Headlights, bumpers and mufflers must be removed. Any make or model closed car can be run, but no roadsters or convertibles are allowed." Old-time true race fans admittedly were shocked.

It was a true jalopy competition, with a total prize of $2,000 set for the feature 15-mile run and six other races. Attracted was a field of 95 cars, but the traffic jam in getting to the track reduced the starters to 40.

The first heat was remarkable in that the winner was Sara Christian of Atlanta, Ga., one of the two women drivers entered. She motored to a close victory in her red and white Ford over Charlie Barry of Syracuse in the 13-car event. Her time for five miles was 4:28.44.

Later Sara finished a creditable fourth in the feature won by Roy Hill of Fitchburg, Mass., over Frank Mundy, a thrill show "Hell Driver." Hill's time for the feature 15 miles, 13:57.52 took auto racing back into the dark ages.

The following Saturday 85,000 persons jammed the grounds for "Race Day" and the resumption of the 100-mile A.A.A. championship. The crowd began arriving at 6 a.m. and by eight had filled every available spot along the rails. Many clung to trees and perched on homemade scaffolding atop trucks, cars and all nearby fair buildings.

The track was a dust bowl despite the efforts of maintenance people in watering it down and applying tons of calcium chloride. Sheets of dirt thrown by the speeding cars covered everything and everyone to windward, which meant the 8,000 in the grandstand.

The race itself almost didn't occur. The drivers, under contract to receive 40 per cent of the grandstand gate, took one look at the infield crowd and decided they'd rather share 40 per cent of the fair's total attendance. After a lengthy meeting which held things up for half an hour, some of the cooler heads among the old-timers pointed out that "we owe it to our fans to put on a show . . . they've been here since 6 o'clock and it's not their fault."

This thinking prevailed and the first run for Indianapolis cars (including a six-wheeler) since 1941 took to the track. At the finish it was Johnny Parsons, 26, and at the time the youngest driver in the field, taking the checker from Starter Duke Dawson.

The New York State Fair was once again off and running.

And it was on this note that Bligh A. Dodds relinquished leadership of the fair to Harold L. "Cap" Creal of Homer, who was appointed director on May 27, 1950. No one could charge Creal with failure to understand the needs of the farmer. He owned an 800-acre spread in Cortland County with a herd of 140 Holsteins. The 53-year-old native of Chautauqua also held a degree in agriculture from Cornell.[10]

The race track again proved a disaster area during the 1953 season. On the first lap, Chuck Stevenson of Milwaukee tried to pass between Jimmy Bryan and the outer wall at the end of the backstretch. He touched Bryan's vehicle, then hurtled over the low concrete wall into a crowd of several hundred persons. Both drivers escaped unhurt, but a spectator was dead and others injured.

It remained for the Labor Day stock car race of 1954 however, to set a new high in ridiculous behavior which survives to this day. A great number of cars were entered in the Eastern States Championships and the 10-lap consolation race was started with 43 bulky cars. As the field reached the second turn, one auto bounded off the outside wall and ricocheted to the inside. Literally everyone piled on.

In what NBC-TV's national news that night called the country's biggest Labor Day wreck, 28 cars were mashed two-deep in places, jamming the track for an hour before tow trucks could extricate all of the drivers and clear away the mess. There were no serious injuries although a number of spectators complained of bruises from a variety of flying bumpers and fenders, or from being knocked down by the stampede to the scene.

A crowd of 448,562 came to the full 1954 fair, with new parking fields on the Solvay waste beds to the north of the grounds packing in an additional 4,000 cars a day. Work on the lot, to eventually handle 16,000 vehicles, was begun in March. Parking officials figured out that 110,000 autos (as against a mere 66,000 the previous year) made it to the fair.

The most popular 1954 display? Eggland. Even as the exhibitors began to disassemble the unit at 2 o'clock on the final day, a large crowd of viewers were enthralled by the "trained" chicken which operated an electronic baseball game. She received a few kernels of corn every time she hit a "home run" by pecking a button activating a bat.[11]

The administration of the fair reverted once again to the Democrats in 1955 following Averell Harriman's election as governor. Although the eight-day 1955 fair under Agriculture Commissioner Daniel J. Carey reached an all-time record of 453,453, the following year bad weather and a new advance sale plan with the 50 cent half-price tickets good only Tuesday through Friday, combined for a severe drop in attendance to 369,864.

The periodic questions about the fair, its reason for being and whether or not it should be a moneymaker, resurfaced during these years, much of it in the form of political rhetoric.

The fortunes of politics found Nelson E. Rockefeller as governor in time for the 1959 fair, and an entirely new administrative team moved in. Harold Creal was back as fair director.

In a way, though, the fair was more broadly a part of the community than ever before. During the 50's every local television and radio station moved operations, lock, stock and barrel from downtown or suburban headquarters to broadcast live from the grounds.

The two Kays of local broadcasting, Kay Larson of WHEN and Kay Russell of WSYR, staked out their locations in the Women's and Pure Foods Buildings for on-the-spot coverage of home demonstrations, cooking highlights and man-and-woman-on-the-street interviews. Frantic electricians and engineers did their best to prevent catastrophes as the massive cameras, heavy cables and lighting standards contrived to trip up anybody passing in the congested areas.

Photographers from out-of-town newspapers found the fair a great place for pictorial features. Many of these were city slickers from the New York metropolitan area, learning for the first time there really was an Upstate.

The *Herald Tribune* sent one of its ace photographers for a feature layout. A city dweller we'll call Ted, he was especially intrigued by the huge cattle show.

A GREAT DEAL OF BULL

Catching sight of a gigantic white bull with great rolling red eyes, the photographer thought a picture of the beast and a youngster would combine the two major elements of a good feature . . . animal and child.

A perky farm girl, perhaps five years old, was discovered nearby and the bull, with two handlers straining on the chain secured through its nose, was soon ensconced in the stable doorway, its great bulk seeming to swell by the second.

The creature's eyes were rolling in even more distressful fashion, and it was making alarming noises as the tiny girl was hoisted close to the ring. The straining biceps stood out on the men's arms.

The New York photographer, armed with an old-time Speed Graphic and flash attachment, inserted a bulb in the unit, looked through his viewer and gave final instructions:

"O.K., I just want Judy and the cow in the picture. You guys, drop the chain and get out of sight."

Although the danger of a snorting, maddened bull careening down crowded fairgrounds streets may have been lost on Ted, it has haunted fair administrators since 1841.

They'd been there . . . on September 7, 1936. That day several handlers were grooming an 18-month-old Hereford steer for exhibit later that afternoon. The 900-pound beast was being hosed down in a wash rack.

It's impossible to guess what thoughts were going through its massive head. Then suddenly the steer took several violent plunges, twisted out of its halter and was running full tilt around a corner of the Coliseum, into the screaming midway crowds.

The handlers were far behind, but other "cowboys" sensing the emergency grabbed at the animal's tail. As the steer was cornered between a couple of parked cars, two state troopers took up precarious positions near the beast's head and horns.

The creature wasn't ready to quit. It lashed out, hitting John F. Doyle of Troop "B" in the leg with its hooves, then lunged away, trampling the policeman as it fled.

A woman wandered into its path and was sent flying.

Back past the Coliseum it went, nearing the front gate before the handlers caught up. An effort to tie it to a turnstile failed when the steer ripped the unit out of the ground.

Other farm hands came up. Finally, the combined efforts quieted the animal.

It quickly paid for its five-minute romp. The next day it became hamburger. Trooper Doyle wound up in the hospital with multiple injuries.

Front gate during 1948 Interim Fair when admission was 30 cents.
Syracuse Newspapers

Coliseum, left, and horse building were focal points for 1948 Interim Fair.
Syracuse Newspapers

Midway in 1951. Coats, shirts and ties were still commonplace dress for fairgoers in those years.
Syracuse Newspapers

Mounted trooper finds plenty of admirers.
Al Edison

Ferris wheels and people blend with sky to provide a mixture of geometry, nature and human reaction.

Al Edison

Display of old horse-drawn rigs attracts Fair officials. Fourth from right is Don Wickham, then-Commissioner of Agriculture; to his right is Mrs. Frank Thompson, while Harold Creal, Fair Director, is at extreme right.

Canal Museum of Syracuse

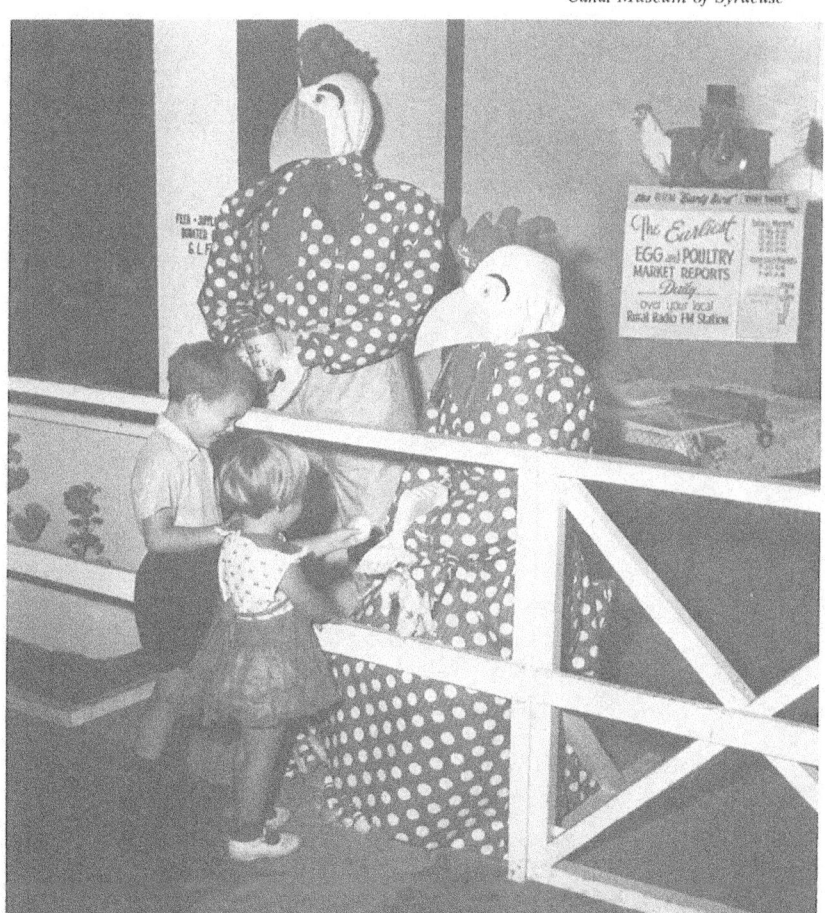

Egged on.

Al Edison

Trotters in action. Note old bleachers and wooden inner rail which prevailed through the 1960's.

Al Edison

Case Steamer, always a crowd pleaser, provided farm power into the 1920's.

Al Edison

Small scale steam locomotive always draws a full load of passengers. Note, they're not all children.

Al Edison

The late Dean Harris of WHEN Radio enjoys an interview.

Al Edison

Action at the duck blind.

Al Edison

Fetch. Retriever takes to the Empire Court lagoon before an appreciative audience.

Al Edison

Six Nation Leadership at the Fair.

Al Edison

Fancy horsework at the State Fair Coliseum.

Al Edison

Uncle Sam inspects the higher sections of waterwheel at Witter Museum
Al Edison

Prize winning roses.
Al Edison

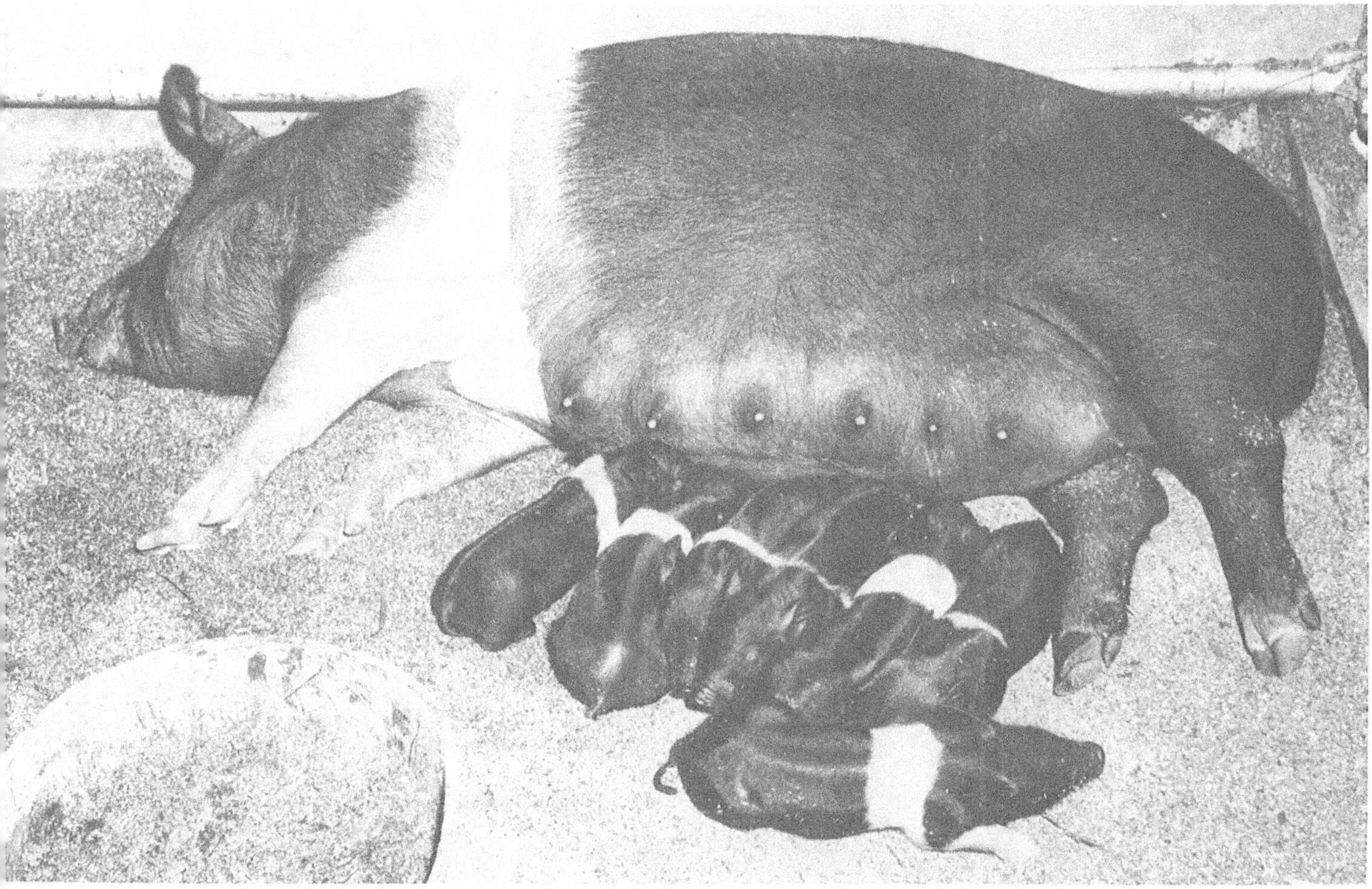

Hunger time.

Al Edison

Giddyap . . . teams strive in horsepulling contest.

Al Edison

The pits in the years before sophisticated high-speed stockcars took to the track. Note narrow tires and preponderance of 1930's coupes.

Al Edison

Jumper on the Coliseum tanbark before full house at Fair's International Show.

Al Edison

Where are the wires? A magician and his helper provide a unique way of opening the Main Gate.

Al Edison

1959 Governor's Day parade featured this Waterloo steam engine made in Canada.

Al Edison

Missile at Fair in 1959 as part of defense preparedness display.

Al Edison

Wooly teammates.

Al Edison

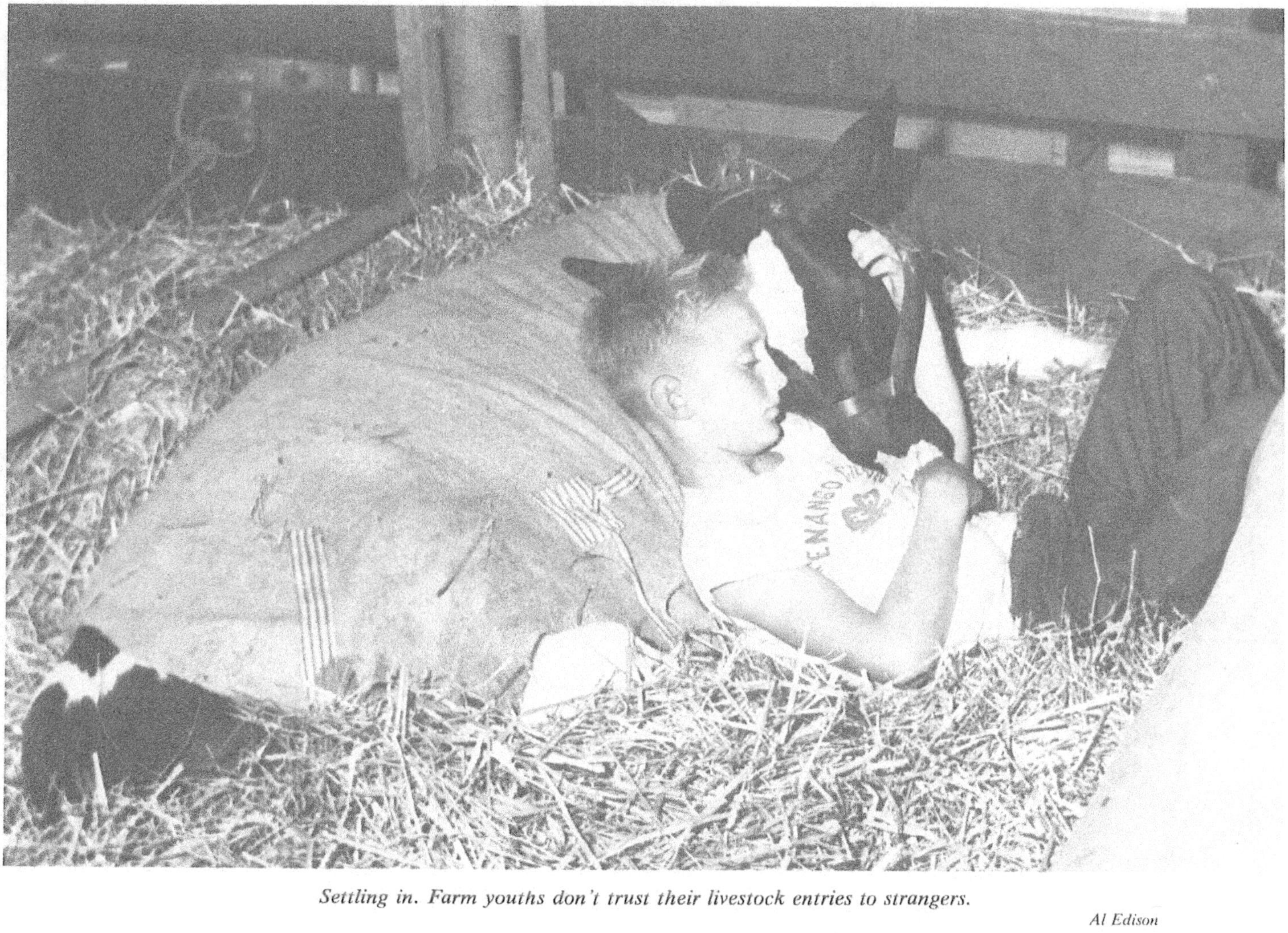

Settling in. Farm youths don't trust their livestock entries to strangers.

Al Edison

Tagged. State police sergeant pins identification tag on young fairgoer, who doesn't seem to mind.
Al Edison

Proud mother and her brood.

Al Edison

Mother, daughter and friend.

Al Edison

Shear fun.

Al Edison

*Here's how. Young William Tell
learns the archer's art from an expert.*
Al Edison

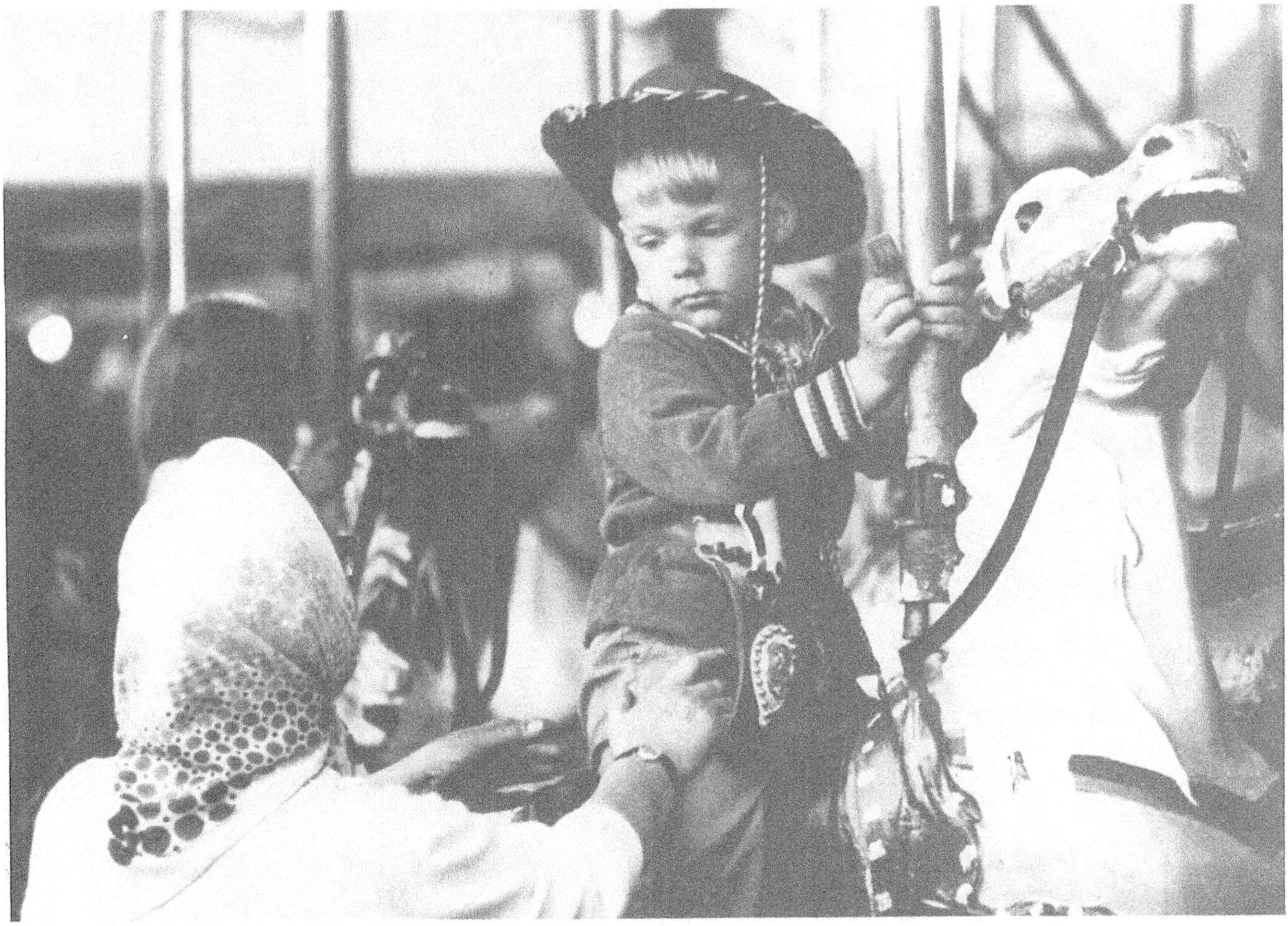

*So much to hang onto . . .
a ticket, pole and reins.*
Al Edison

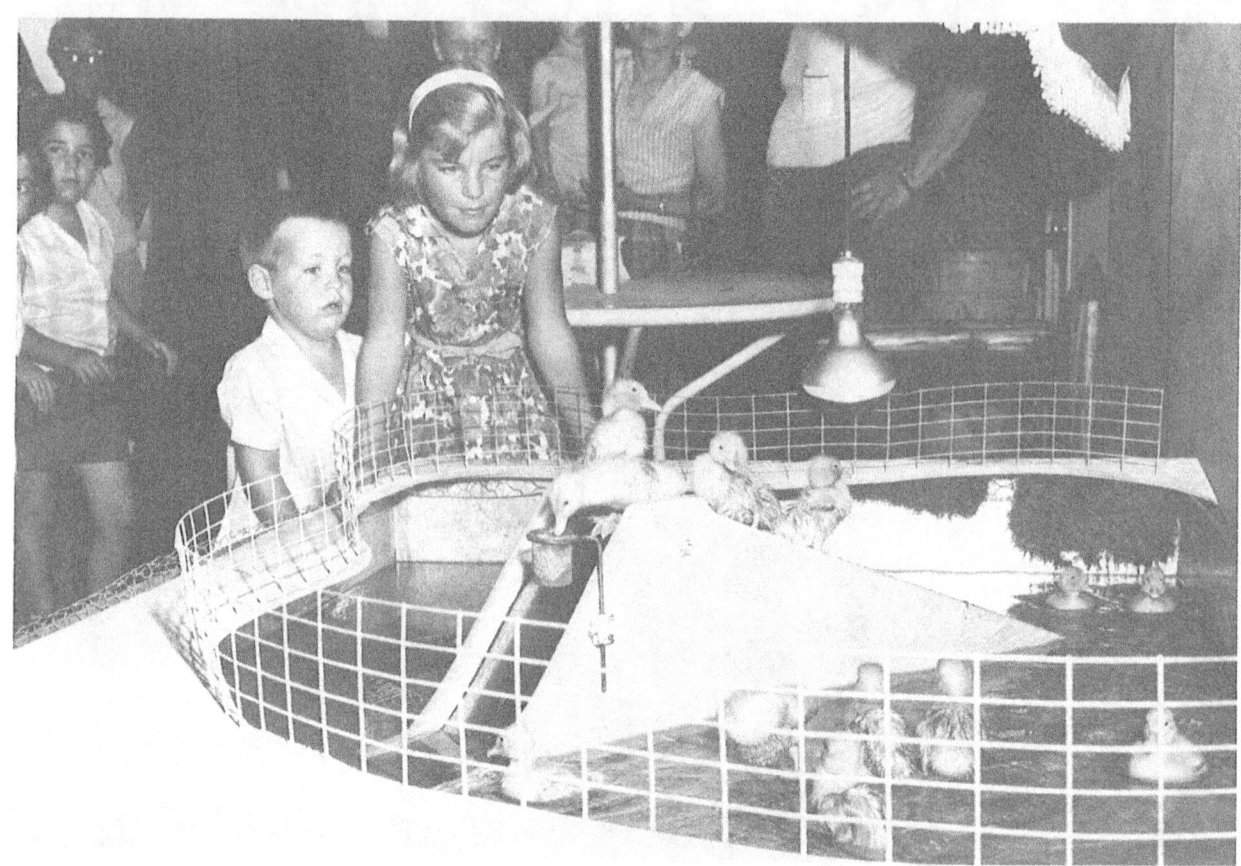

Just ducky. The slide proves popular for those in and out of the tank.

Al Edison

Puppy love.

Al Edison

Wall to wall junk. How many racers can you count in this back-stretch jumble?

Al Edison

Tennessee Ernie Ford drives a namesake tractor.

Al Edison

Just testing. Pie competition draws an admirer.

Al Edison

Guy Lombardo, the bandleader, and friends.

Al Edison

CHAPTER VI

The Modern Era

The space age, the Vietnam War, disturbances on campus and on city streets and the rampant inflation of the 1970's all had their effects on the Syracuse area and in turn on the state fair.

The space age truly arrived at the 113th fair on September 4, 1959 being accomodated by the earliest official starting hour in the exhibition's history . . . an eye-opening 7:15 a.m. It was all duly scientific, because that's the moment when the Vanguard I Satellite passed overhead on its 5,759th orbit.

At precisely that time, as E.F. Herzog, manager of engineering at General Electric Company completed the countdown, a signal from the satellite was picked up by a "magic wand" held by State Fair Queen Carol Youmans of Schenectady.

The impulse activated motors in the main gate, causing the 50-year-old entranceway doors to swing open.

The fair was truly a combination of the old and new, with antique farm machinery such as an 1866 horsedrawn steam engine from the Ames Iron Works in Oswego to a ballistic missile erected in Empire Court, and demonstrations of atomic power in the Industrial Building.[1]

A "new" first, free entertainment at the grandstand featured the talents of Edgar Bergen and Charlie McCarthy, Art Linkletter, Guy Lombardo, Tommy Sands, Johnny Cash and Jay P. Morgan.[2]

The 1960 fair went well over the half-million mark, with 507,462 going through the turnstiles. But the following year the shadow of a half-centruy earlier again darkened the grounds as a polio scare caused attendance to drop by 64,000 to 443,655.

The 1961 event was remarkable for the tremendous display of historic farm machinery and equipment, the showing named in memory of Williard J. Durkee, a Syracusan who'd served as superintendent of the fair's first antique farm equipment display two years before. The fair exhibition, under the direction of Lester Norris of Marcellus, then-vice president of the New York State Steam Engine Association, was set up in a lot in the western end of the grounds.[3]

A large number of steamers were displayed in working order, their plumes of black smoke enveloping the area in pungent haze, cinders, occasional red-hot embers and chugging, pumping and wheezing sounds. The units in shining black, red, yellow and green with glistening brasswork and huge iron wheels, were equipped with Baker fans and 100-foot belt power transfers to operate threshing and sawing equipment.

The ghosts of Buffalo's 1857 fair lived once again for a day.

141

"EXPO"

The fair was in for a number of changes in 1962, with the State Legislature opening the action in February when it repeated history by again switching the show's name from "fair" to "exposition." This of course was in keeping with a part of Governor Hughes' 50-year-old injunction wheeled out by proponents to justify the change . . . even though it hadn't worked in '38.[4]

Despite the "million dollar farm machinery show," the "Spotlight on Construction in the Empire State," the theme of the 1962 fair and the name "Manufacturing and Liberal Arts" on the ground's largest permanent structure, most "fairgoers" found it difficult to swallow the idea their state fair was equally agricultural and manufacturing and that the new name was appropriate.

Veteran fairgoers such as 88-year-old Joe Richel who had been an attendee for 54 years said, "The fair will always be the fair." He had another, more personal concern. Living at 1116 State Fair Boulevard, he asked, "Will they change my address to State Exposition Boulevard?"

"It sounds corny," said another fair fan.

State Senator Laurence M. Rulison who introduced the bill in the legislature at the request of Harold Creal, said he took the action because he believed the fair needed "a new image both physically and other ways. A fair denotes agriculture; and exposition offers more."[5]

In truth there was plenty more, at least to the name. The full title was "The Agricultural-Industrial Exposition of the New York State Department of Agriculture and Markets."

Signs around the grounds were changed. The word "fair" was obliterated from stationery, advertisements and publicity.

The change did stir up some alarm amongst farmers, and Fair Director Creal quickly explained, "Agriculture would not take a back seat," and the the "exposition" would still have cows, pigs, pumpkins and squashes and "cute ducklings coming down a slide."

But most importantly, he told the farmers $150,080 in prize money awaited exhibitors, $138,000 going for agricultural produce and events.

After taking a long look at the attendance situation, it was decided by state authorities that one of the problems was starting the event too late, with most youngsters back in school the day after Labor Day. Consequently, the Expo's 1962 schedule commenced the last week in August, closing Labor Day night.

This in turn left the 100-mile championship auto race without a date during the fair even though the event was scheduled for the usual Saturday after Labor Day. So, when the United States Auto Club event (featuring among others, a young upstart from Texas named A.J. Foyt) received the green flag, instead of the usual packed stands and infield, a skimpy 3,400 were in attendance. The state lost $12,000 on the race and after failure to find a promoter, it was dropped.

THE MAKING OF A BLOOD BROTHER

The fair's director himself became part of the show the next year when Harold Creal was made a blood brother of the Iroquois Confederacy, thus joining such eminent predecessors as Onondaga County's pioneer settler and Revolutionary War hero, Ephraim Webster, and

James Geddes, the surveyor of the Erie Canal.

Indian chiefs of the various nations, among them the Cayugas, the Mohawks, Oneidas, Onondagas, Senecas and Tuscaroras, garbed in buckskins, bright ornaments and feathers stood in line, chanting the ancient ritual in their native language as Creal was escorted past them in a symbolic running of the gauntlet. The leaders stood in a semicircle on the ceremonial mound as their new blood brother stepped slowly past each member, standing face to face for a brief introduction.

Mrs. Rita Peters of the Onondagas was named clan mother for Creal, who was accepted into the Turtle Clan of the Onondagas by a vote of group elders, and given the name, Tan-Noh-Doonk, meaning the "Head One."[6]

Several years later Creal retired as director, being succeeded by Bernard Potter of Truxton, one of his Cortland County neighbors. He left on an upbeat note, with attendance in 1966 a record 525,000 and the Manufacturers and Liberal Arts Building renamed Center of Living Building.

One Creal innovation however, didn't last. Local Assemblyman John H. Terry and State Senator Tarky Lombardi, Jr., co-sponsored a bill which officially changed "Expo" back to "State Fair". They had a strong supporter in Governor Rockefeller who expressed the opinion it would be nice to once again have a "State Fair."[7]

Besides, 1967 was the year for Expo '67 in Montreal and no one wanted things confused.

Completion of the Route 81 Interchange through Syracuse coupled with local Route 690 meant a tremendous saving in time for tens of thousands of Central and Northern New Yorkers and Canadians enroute to the fair.[8]

The event certainly couldn't be charged with being oblivious to progress. On August 26, 1969 a "spaceman" from the Bell Aerosystems Demonstration Team flew over the front gate in his rocketbelt to open the event. And a chunk of moon rock was displayed, courtesy of the National Aeronautics and Space Administration.

The theme? Appropriately enough, a salute to the "Space Age."[9]

THE SEVENTIES

The decade of the Seventies was marked once more by changes in state fair policies, a switch again to Democratic control and record attendances. The one-day mark fell in 1970 as 113,795 persons crammed the grounds to snap the old record achieved in 1967 by 2,300.

But it was also a year of fatalities as yet another person, a spectator, died and six were hurt when a stock car jumped the fence at the same old spot on the third turn, landing in a crowd watching from a restricted area. Two pedestrians, a carnival worker and a young woman, were fatally injured during the first two days of the 1970 fair when they tried to cross Route 690 near the Main Gate.

THE DEMOCRATS TAKE OVER

The Democrats once again took over the reins in 1975 following Hugh Carey's victory over then-Governor Malcolm Wilson the previous fall.

On May 2 Thomas Young, the 27-year-old son of the late Syracuse Democratic County Chairman, John "Bocko" Young, succeeded Norman Rothschild as fair director. The new director agreed with Rothschild on one basic point: it was essential to get year-round usage out of the buildings and grounds in order to warrant the growing upkeep costs. (In 1982 alone fair improvements cost well over $3 million).

Inflation also had its effect. Where gate admission prices had gone to $2 in 1971 from $1.50 the previous year, advance sale prices were upped to $1.25 and parking increased to a dollar from 75 cents. (Four years later admission was $2.50 and advance sales at $1.75. By 1982 it had gone to $3, and $2 for those buying ahead of time).

Sometimes it takes a little while to get untracked, and that's what happened in 1975, when opening ceremonies just didn't seem to want to happen. While 15,000 persons waited patiently for the 9:30 a.m. start, 60 band members waited impatiently for the truck with their uniforms.

The parade finally got underway an hour late. But then the procession had to stop at Empire Court because a rock group was on stage setting up sound equipment.

When the speaking program finally started, Director Young had to contend with an 80-piece military band which took that moment to leave and the simultaneous resumption of work by the rock group.

It was then Agricultural Commissioner John Dyson's turn, and he invited fairgoers to be sure to buy their tickets for "The Great American Music Festival." It was also his introduction to music at the fairgrounds, an indoctrination which was to get a lot louder before it got better.[10]

ROCKY TIMES

The fair, through its State Industrial Exhibit Authority, had authorized rock concert promoter John Scher to put on a day-long show on September 2, the day after the fair itself closed. To be featured were the Jefferson Starship, the Doobie Brothers, Beach Boys, New Riders of the Purple Sage, America and the Stanky Brown Group. A crowd of 100,000 was anticipated.

Things didn't get off to a rousing start when the grounds, including the infield, was turned into a mudbowl by a 24-hour downpour. Even so, when the gates opened at 6 a.m. the crowd of fans began to peacefully file into the soggy grounds for the noon concert.

By 8:15 about 8,000 persons were there. Most were part of an earlier scene when thousands stayed up all night to hear a local rock band on the hill between the parking lot and Onondaga Lake. Once ensconsed in the infield mud, ankle deep in spots, many wrapped themselves in plastic and went to sleep. As an added attraction, a huge frisbee game took shape with something like 100 participants.

Dope was present too, and several booths were doing a thriving business in marijuana paraphernalia.

Then things got testy. Close to a thousand would-be concert goers not necessarily interested in paying their ways in, converged on the gate. A sudden barrage of bottles and beer cans followed.

The state police were ordered to back up, as were the promoter's security guards. An estimated 30 to 40 security dogs went into action. The mob kept trying to push through despite Fido, and were pulling the gates off the hinges when a fairgrounds fire truck pulled up, hosing down the crowd. That ended it.

In order to keep the crowd (estimated at close to 40,000 by noon) content, the concert started at 11 a.m. By the following day all that was left was a pile of trash and a whopping big unpaid bill.[11, 12]

Dyson, never at a loss for a pithy comment, said of the concert:

"It was not a success.

"It was an unsuccessful enterprise.

"It was our last rock concert."[13]

A PERIOD OF GROWTH

Anxious to build up the gate, the administration turned the clock back a century in 1978 as the fair was kicked off by a balloon ascension. The multi-colored General Electric hot air bag, safely tethered by a 50-foot rope, lifted slowly off with a basketful of fair officials. It gave Tom Young the cue to say, "I am high on New York State and on the New York State Fair."[14]

As a way of avoiding the terrible traffic congestion within the grounds, the Main Gate for the first time in history was closed to cars and trucks. Automotive movement in the heart of the grounds was also limited . . . to golf carts.

And, while the big car 100-mile USAC races may have left the fair scene, another form of championship auto racing has grown in popularity . . . the state microd races on a tenth of a mile paved track tucked away in the northwestern corner of the grounds. More than a hundred microds, homemade cars with lawnmower engines, operate in various classes with boys and girls ranging in ages from 5 to 17 behind the wheels.

The conflict between breaking even and running a fair resulted in the elimination of the free grandstand shows in 1976. Thus the Bob Hope, Charley Pride, Neil Sedaka and Bobby Vinton appearances of the Bi-Centennial year were switched to the grandstand with special admission charged . . . a reversion to the practice of the mid-1950's.[15]

While perhaps not enough to balance the overall budget, it's still big business, with the fair collecting a gross of $122,780 in 1980 for the one-day appearance of the Charlie Daniels Band, an all-time high to that point. Attendance for the evening was 15,854. Under the old procedure instead of a profit, the fair would have been in the red for more than $100,000. (The Daniels record was shattered in 1982 when 15,954 attended a Willie Nelson concert).[16]

The fair is big business by any measuring stick, with an annual budget of $3.5 million and a year-round staff of 50 persons. Some 300 food contractors and concessionaires provide the pizza and sausages, doughnuts, candied apples, pop corn and cotton candy. More than 3,000 persons are listed each year as exhibitors.

Close to a quarter of a million dollars is distributed to the 20,000 top entries in events ranging from flowers to cheeses to hogs and poultry and to owners of some 3,000 head of cattle, making the show one of the world's biggest livestock gatherings.

The fair fills 4,200 local hotel and motel rooms for the 10-day annual spree, while taxes generate an added $160,000 income for Onondaga County. Some 500,000 gallons of gasoline are also needed to accomodate event attendees.

Cumulative benefits to the Syracuse area are estimated at $42 million each year![17]

The physical limitations of the grounds has given rise to a gigantic auxiliary attraction held a few weeks before the fair as farmers seek to actually see the giant cultivators and other sophisticated farm machinery in action. Thus the annual Empire Farm Days, held for many years on a hilltop just southeast of Syracuse (and now near Ithaca) draws close to 200,000 persons each year . . . a direct result of the interest in agricultural activities spawned by the fair.

RAMBLING THE GROUNDS

A trip to the fair today, as in the past, can be as expensive or as reasonable as the visitor wants to make it, once the admission is paid. (The average adult, however, according to a 1981 survey spends $22 during a six-hour visit). All you have to do is mix a little fair knowhow with patience, because there will be others in lines . . . often long lines . . . to get such goodies as the baked potatoes served continuously at the Horticulture Building. Prepare to wait up to 45 minutes for one of the spuds.

Other taste treats include smoked beef sausages, a variety of cheeses and accompanying hot mustard, and the foodstuffs at the Youth Bulding where 4-H members offer substantially heartier fare such as hot sauerkraut, various desserts and the specialty, carrot cakes.

Yogurt, dried fruit and nuts, bite-size fudge pieces, cookies and, through the courtesy of the New York Maple Producers Association, a lick of maple cream can also be had for free.

To wash it down, strawberry flavored milk can often be found in tiny dixie cups; a sample of American, French-American and European wines is available at the New York State Wine Grape Growers booth, and there are always the cups of cold water provided thirsty fairgoers at the Onondaga County Water Authority exhibit in the Center of Progress Building.

If you're a bit under the weather, there are plenty of booths where blood pressure is checked and where diabetes tests, a posture profile and similar health examinations are offered at no cost.[18]

And there are the traditional booklets, leaflets, flyers, brochures, pamphlets and tracts on almost every conceivable subject from growing Christmas trees to growing old. They all fit into the giveaway plastic bags suitably inscribed by advertisers . . . standard fair visitor equipment. (Reporters and others have tried to track down these various items and their collectors once they leave the fair, but they both seem to disappear, with the collectors presenting themselves once again at the following year's festivities).

But perhaps you're willing to pay your own way. Then be prepared to shell out up to $1.85 for an Italian sausage sandwich. Polish kielbasa goes for $1.30 while chocolate coated frozen bananas sprinkled with peanuts sell for 75 cents. Lemons are crushed before the buyer's eyes as lemonade is made to order at 75 cents. The New York State Grange roast sandwich booth features beeves being cooked in a very special and aromatic way. Plain, the slices from the beast which has been roasted whole on a spit, go for $2 while barbecued sandwiches are $1.50.

Then, something different is "The Dancing Camel," a sandwich made into a pouch which contains up to and including (at your risk) tacos, cheddar cheese and mushrooms, sour cream and chives and vegetables.[19]

THE MIDWAY MILE

How does the Midway and its rides today compare to the 1849 Ferris Wheel? The city desk at *The Post-Standard* sent reporter Mark Hass to find out, armed with a few dollars and some dramamine.

The Midway, he recalled, is a place halfway, "But halfway to what?"

"For some, even the mention of scramblers, skywheels and superloops scares up visions of being halfway to Hades. Setting foot on a carnival ride is pure hell.

"Then there are those blessed with solid stomachs and steady nerves. For them, the smell of diesel-powered generators, the sounds of blaring music and the silhouette of a Ferris Wheel against an afternoon sky is a scene from a place halfway to heaven. A front seat on a roller coaster is bliss."

Talking to "Captain Zip," the operator of the Astroliner, a newer ride, Hass finally got his answer to why some endlessly test their nerves on rides while others are repulsed by the thought.

"'Some like them, I guess,' he said, 'and some just don't.'"[20]

The James E. Strates Show, which has for many years been THE Midway for this fair and 16 others throughout the east, came into being 60 years ago when the elder Strates and two friends, Nick Bozinis and Bill Platt from the Elmira area, formed a wrestling show for a fair in Bath, N.Y.

They saved their money and bought a merry-go-round and two kiddie rides. As the operation grew, Strates bought out his partners. His son joined the carnival after graduating from Syracuse University and a hitch in the marines, taking over the operation when his father died.

The show, a $7 million-a-year operation with winter quarters in Orlando, Fla., has a special train and 350 employees needed to set up, run and break down the 105 midway rides, games and exhibitions, all of which at night turn into a fairyland of multi-colored neons in motion.[21]

THE BOTTOM LINE

Some of the people who make a fair a fair are more unusual than others.

One of the more unique individuals is Eliseo Rossi of Utica who bicycles the 52 miles each way to be first at the gate when the fair opens. So far, nothing too unusual. But Eliseo in 1984 was 86 years old, and he's been doing this since 1949.

His bicycling since then totaled something like 210,000 miles, including a 1949 round trip to California and a 1953 visit (with the help of train and plane) to the 50 state capitals.

Someone in Syracuse back in 1975 stole his familiar 34-year-old one-speed Schwinn. Disconsolate, he took the bus to the fair for a couple of days, but then happened to spot his cycle in the hands of a 23-year-old city man at a downtown intersection. Rossi regained his bike and the man got three months plus probation, but the Utican has a scar on the bridge of his nose as a souvenir of the encounter.

Although he's been the first through the turnstiles on 27 of the 35 years he's been at the fair, that's not enough. Rossi also holds another distinction. He's usually the final customer to leave after Labor Day.[22]

And he admits that the road to Syracuse can sometimes seem all up hill. In 1984 it took nine hours for the journey, which included a luncheon stop and fighting a bothersome headwind.

What it's all about. New York State Exposition ribbons and trophies surround the 1963 Queen.

Al Edison

Antique and sports cars undergo close scrutiny.

Al Edison

Legendary Colgate Football Coach Andy Kerr becomes a blood brother.

Onondaga Historical Association

Clearing the hurdle in the International Horse Show.

Al Edison

Massed choirs from local churches at 'Pause Before God' in the Coliseum, a Sunday afternoon Fair tradition.

Al Edison

No wooden nickels. Edgar Bergen and Charlie McCarthy were popular entertainers.

Al Edison

Some of the competitors still wear their fenders and headlight sockets, while others are stripped down for action.
Al Edison

Watch it. Fans head for cover during the late '50's as car tests backstretch wall.
Al Edison

Bobby Kennedy campaigning in Empire Court.

Al Edison

The Midway in 1966.

Syracuse Newspapers

I wonder if it bites? Governor Nelson Rockefeller and his wife, Happy, visit the exhibits.

Al Edison

The mighty Wurlitzer from Syracuse's RKO Keiths is now a permanent fixture in the Arts and Home Building.
Al Edison

Two-mule team from Death Valley makes it to the Fair enroute to Montreal's Expo '67.
Al Edison

Civil Air Patrol cadets view Apollo 12 Moon Rock.

Al Edison

Not the highest, but these riders appear to be having fun.

Syracuse Newspapers

The midway takes on a magical aura after nightfall.

Syracuse Newspapers

Microds in action on tenth of a mile paved track.

Al Edison

Square dance spectacular at the Coliseum floor draws a large crowd.

Al Edison

Modified stock cars in mid-60's reveal technical improvements.

Al Edison

The late Frank Thompson, who directed the Witter Agricultural Museum, looks over an antique iron plow.

Al Edison

Flight time, Buck Rogers style, as Bell Aerosystems test pilot demonstrates flying belt to open Fair.

Al Edison

He clears ground.

Al Edison

Up, up and away . . .

Al Edison

Gate crashers storm fence at Music Fair in 1975.
Dick Blume, Syracuse Newspapers

Massive tractor strains in pulling contest, a new version of an old game.

Eliseo Rossi, perennially the first in line at the Main Gate, bicycled in from Utica despite being in his mid-80's.
Al Edison

Microds, homemade racers with lawnmower engines, ready to go on tenth of a mile track.
Al Edison

Haymaker. Fair provides plenty of jobs of all kinds.

Al Edison

Here's the business end of one of the traditionally long lines at the Fair. Baked potato and a dollop of butter constitute the most popular freebie at the annual event.

Al Edison

The eyes have it at the rainbow trout tank in the Conservation Department exhibit.

Al Edison

Fire leveled Farm Machinery Building in minutes. All firemen could do was hose down the ruins.

Carl Single, Syracuse Newspapers

The mile oval is lighted for the July 3, 1984 races, first time the raceway has been permanently lighted.

Lori DuBois, Speedway Press

Modified racers in the Schaefer "200" cost in excess of $20,000 each, and are timed in the 120 mph bracket.

Speedway Press

CHAPTER VII

Off-Season

The year-round aspect of the fairgrounds, a dream of state officials for more than 90 years, is never more apparent than when the Center of Progress Building hosts such annual events as "Gem World," or a statewide conclave of bottle collectors' clubs, a recreational vehicle show or the Northeast Sportsman extravaganza.

More recently model railroaders and airplane fans have held major exhibitions which attracted thousands to the grounds. One recent month proved an excellent example. Thirty-one distinct events were held including the Empire State All-Arabian Horse Show; a miniature doll and furniture exhibit by the Syracuse Area Miniature Enthusiasts; a cheese manufacturing seminar staged by the State Department of Agriculture and Markets and a variety of food demonstrations and antique displays.

The huge grandstand rocks as well, to the thundering of souped-up farm machines in the annual tractor pull with modified and supermodified monsters struggling to drag loads of up to 25 tons along the cinder track.

The 66,900-square foot Youth Activities Center (formerly the Farm Machinery Building) served as the permanent home of the Central New York State Modern Style Flea Market until it was destroyed by fire on April 19, 1983. Every Saturday and Sunday, except during fair week, it offered everything from new and used clothes to foods and stamps, coins and other collectibles in what was termed by those who had never been to New York's East Side, "a carnival New York City Delancey Street atmosphere."

In the summer of 1980 acres and acres of instant city sprang up over the grounds, the infield and the many surrounding parking areas, as 4,200 camping units and 12,600 campers formed a "mobile metropolis" for the annual National Campers' and Hikers' Association convention.[1]

A few years earlier the year-round scope of the fair was enhanced with the dedication of the Empire State Theater and Musical Instrument Museum in the Art and Home Center. Three years later, in 1970, the mortgage was burned to the music of Karl Cole on the former R.K.O. Keiths Wurlitzer Organ.

As befits a fair, cattle shows and auctions are not limited to fairtime. The grounds on June 23, 1980 was the scene of the sale of a Holstein cow owned by the late John Lennon and his wife Yoko Ono, for a reported world record $265,000. The cow, bought by Steve Potter of Martha's Vineyard, was expected to produce 50,000 pounds of milk annually.[2]

BRING BACK THE HAMBLETONIAN

Ever since the Hambletonian left the fair for Goshen, efforts have continued to bring it back to its birthplace. But in 1957 when Goshen could no longer provide the modern environment and facilities, instead of returning to Syracuse the race was awarded to DuQuoin, home of the Illinois State Fair.

In a way this was understandable. The post-war Syracuse fair was not really oriented toward harness racing, especially since nearby trotting tracks worked hard to prevent the fair from taking away their livelihood. Consequently the Grand Circuit never did reappear; the rustic bleachers by the railroad tracks were ill-suited to the more sophisticated horse race attendee, and there was no pari-mutuel.

This lack of gambling proved one thing conclusively. Even though annual races were held in conjunction with the fair and even though many of the country's best trotters and pacers were matched in races with purses exceeding $300,000, the events were often run before crowds of 500 or less.

DuQuoin, which offered perfectly manicured grounds with a parklike landscaped infield, also had its limitations. It too did not encourage gambling. The summer heat, lack of air transportation and limited hotel-motel accomodations added to the problems, so that by 1972 attendance for the nation's premier light harness racing event had slipped to just 8,000.

New Yorkers combined their talents in trying to retrieve the race. Donald J. Wickham, then-Commissioner of Agriculture, had several years earlier urged in a letter to Governor Rockefeller approval of funds for an adequate grandstand to assure the Hambletonian's return. He anticipated attendance of from 30,000 to 50,000 persons plus national television.[3]

The governor responded on October 21, 1970, pledging his full support for a new grandstand saying, "I recognize the need to provide the kind of facilities that will ensure the return of the Hambletonian to New York State."[4]

Despite intensive local efforts by the Syracuse Chamber, Mayor Lee Alexander and County Executive John Mulroy and added support from Governor Malcolm Wilson, Rockefeller's successor, the Society apparently became caught up in a glow of patriotism and went for a three-year agreement with the Liberty Bell Park in Philadelphia covering the Bicentennial period. The Park however, backed out of the project and the race stayed in Illinois.[5]

By 1979 it was once again pretty well decided to move from DuQuoin, and Syracuse interests produced a tempting package. The new $3.5 million dollar, 15,000 seat grandstand was a reality. Pari-mutuels had been installed and broken in during races on the newly-designated Syracuse Mile. (The "Mile" was the result of a sustained push by Syracuse interests to develop a meet equally as important as the Hambletonian. Norm Rothschild, then fair director, and officials of the Harness Breeders' Association put together the final program).

Rivalries between the fair and local trotting tracks had been amicably worked out so the full support of harness racing promoters and horse owners could be expected. But it was thought the convincer was money. New York horsemen and trotting tracks pledged $450,000 annually.

The New York Legislature agreed to create a public benefit corporation to manage a Hambletonian Events Fund to receive a percentage of the state's pari-mutuel tax revenues and Hambletonian bet income. Also guaranteed was a minimum of $3.6 milion in purses in an added-event 10-day meet.

The Hambletonian directors met in Lexington on October 5, 1979. And they voted down the $3.6 million New York State bid in favor of sending the race to the Meadowlands in New Jersey, even though New York guaranteed $1 million for 1981 against an $800,000 Meadowlands offer.

The directors' votes among Meadowlands, DuQuoin and Syracuse went: Meadowlands

9, DuQuoin 8 and Syracuse 4. The second vote with Syracuse eliminated, was 21 for Meadowlands, 12 for DuQuoin. Gaston Valiquette, general manager of the Syracuse Mile Meet, told *The Herald-Journal*, "I do not feel the decision was fair and I don't believe the vote was on its merits . . . DuQuoin may decide not to have any racing at all and certainly the future of the Zweig Memorial trot and racing at the Syracuse Mile is in doubt."[6]

To say that the mile-long fair track no longer had much to offer the harness racing set was disproved on the balmy Saturday afternoon of August 16, 1980 when superhorse Niatross outraced a field of four others in setting an new all-time harness record for the mile of 1:52.4. It was a fifth of a second under the old world mark.

Weather conditions, according to many horsemen, were not exactly perfect, with the temperature at 73 degrees and a northwest wind coming in at 14 miles an hour, blowing against the horses in the backstretch and three-quarter turn. Clint Galbraith, the driver-trainer, believes the wind slowed Niatross by at least a second.

Additionally, the field of Volos Mike, Miles End Brenda, Bill's Advice and Timely Fella were not up to pushing Niatross in the non-betting event.

Then the horses took to the track.

Jack Bailey drove Timely Fella hard to a close quarter in 27 seconds and stayed in sight to the half at 55.3, but then Niatross moved out, hitting the ¾ pole in 1:23.4 and finished 21 lengths ahead of Volos Mike.

It was a popular win for the $10 million colt and Gailbraith was offered a $250,000 bonus by the owners of the Niatross syndicate for establishing a new world mark.[7]

FAIR ON ICE

Winter sports found a convivial home in the Coliseum, commencing in 1930 when an International Hockey League membership was awarded the city after a group of promoters visited Syracuse and decided the Coliseum was the most suitable site.

The first Coliseum team took over the Hamilton, Ontario franchise while Toronto was replaced by Pittsburgh. Others in the league were Cleveland, Buffalo, London, Ont., Detroit, Windsor, Ont., and Niagara Falls.

The poorly financed Syracuse squad was a classic of ineptness, and was described as being "just as able in a barroom as on the ice." It started last and stayed there. The sheriff was often present at the Coliseum box office to take care of the gate receipts for anxious creditors.

The next year refinancing enabled the team to make a comeback and the city was a hockey town until the war years. In 1936 the American Hockey League formed with the Stars one of the original eight teams. Led by Jack Markle who took the league's scoring honors that year, the sextet won the championship.[8, 9]

After the war there was a dearth of hockey at the Coliseum until the early 1960's when the resurrected "Stars," an amateur squad of Canadians, played every Saturday and Sunday against visiting Senior "A" teams from Quebec and Ontario.

With the Stars, a new crop of yellow and black clad favorites . . . the Crawford Brothers, Lefty Fregin, Wiener Brown, Goalie Norm Parrish and Coach Gogi Goegan . . . took to the ice, appearing against such teams as the United States Olympic Squad, the Russian National sextet and the Rochester Americans of the American Hockey League.

It was two days before Christmas in 1964 when the Stars and the Russians squared off before an overflow house. The Stars were not a mere pickup team. Many of them played years before with the Belleville MacFarlands when they engaged Russia as members of the Canadian World Cup team, and were fully aware of standard Red strategy.

Opening with a heavy offensive, the Russians put pressure on Goalie Parrish, but it was the Star's Bill Crawford who took a pass from brother Frank to notch the first goal at 3:52. By the end of the second period, Syracuse had a 5-3 lead following three successive goals: a power play counter by Bill Crawford; Claude Charron's 30-footer and a Weiner Brown rebound.

When the teams came out for the final session, the knowledgeable fans noticed a difference. Some of the bulky civilians sitting behind the bench were no longer there. Instead, they were in Russian uniform.

Syracuse's limited reservoir of players couldn't hold out against the fresh troops, and went down to an honorable 7-5 loss. (The Soviets, incidentally, were within their rights. According to international hockey rules at the time, a team could carry extra players who could be exchanged for others).[10]

The amateurs couldn't survive competition from the newly-arrived Blazers of the Eastern Hockey League, then playing in the downtown War Memorial. Shortly afterward the Coliseum became a haven for the Midstate Youth Hockey Association.

The youth sport during these years was taking off. Not just one, but two rinks were set up, one in the Coliseum proper, the other in an annex built to the west. Both are busy from 7 a.m. until midnight seven days a week during the winter season for junior, high school, industrial league and the Syracuse University Club teams.

HOSTING THE ORANGE AND THE NATS

Basketball moved into the building for the 1947-48 campaign, the year after a major fire at Archbold Gymnasium left the Syracuse University team without a campus court. Downtown's Jefferson Street Armory became a substitute site but you couldn't get more than 3,000 people in, and it was recognized that a lot more than that wanted to see the Orange cage team with scrappy star Billy Gabor in his senior year (he was to lead the team in scoring from 1945 to 1948 with 1,344 career points).

Consequently the cavernous Coliseum, all by itself in the deserted fairgrounds was checked out. But there was no basketball court.

Lewis P. Andreas, then the coach, contracted for a magnificent, wondrously heavy wooden floor made from maple and moveable to the point where the court could be lifted for annual horse shows and livestock events, but not every other day. The cost . . . $10,000.

Fans took to the fairgrounds games. The experiment was so successful that scholastic games were soon being booked for dates when the Orange wasn't at home. Eventually the Syracuse Nats of the National Basketball Association sought the Coliseum for its matches, too.[11, 12]

Meanwhile, ever a sharp horse trader, Andreas took the opportunity to sell the state on acquiring the court . . . for $10,000.

The Nats found the years at the Coliseum some of the happiest of their stay in Syracuse. It was the time of Billy Gabor, Big Ed Peterson and Paul Seymour, and heroic games which rewrote pro basketball's record book.

Such was the Thanksgiving Day game in 1949 which has been described as "the greatest basketball game ever played." At least it is so remembered by the 6,821 fans, a sellout house which crammed into every corner (including seats behind the posts). Four hours after they'd taken their seats they left, absolutely drained of emotion.

Syracuse's opposition that night was the Anderson Packers, the defending league champion. And for the regulation 48 minutes, the teams finished 76-all. The Packers moved to a slight lead just before the first five minute overtime came to an end, but Ray Corley's free throw made it 83-83.

The second overtime was a ball control effort. Ed Stanczak of Anderson dropped in a foul shot with 12 seconds left to erase a Syracuse lead. Score: 87-87. Sixteen seconds remained in the third overtime when Wally Kirk of Anderson cooly sank two free throws to tie up the score, while in the fourth stanza Fuzzy LaVane dropped in a "swish" set shot to knot the score at 107 each.

Then, in the fifth overtime with almost all players using up their final fouls, Paul Seymour and Ray Corley each sank three straight free throws — just enough to stave off the still-driving Packers. Stars for Syracuse included Johnny Macknowski with 21 points and Dolph Schayes, Paul Seymour and George Katkovics, each with 18.

But what was most remarkable is that the game set all-time National Basketball League records of 125 points for one team; 248 for two teams; the record for most fouls called, 123; the most free throws made; most free throws missed, most overtimes and longest major league game (73 minutes). For many years it was pro basketball's highwater mark, and is still the highest before the 24-second rule was introduced into the professional game.[13, 14, 15]

Danny Biasone, president of the club, was happy about the Coliseum arrangements for another reason. Many of the seats at the Coliseum were behind posts, meaning that patrons bought their tickets well in advance to assure a good field of vision . . . an advance sale which is especially important considering Syracuse winters and the temptation for fans to stay home on a bad night. (When the team went into the newly-completed War Memorial downtown for the 1952-53 season, this habit carried over for another season before fans realized almost any seat was good, and advance sales dipped substantially).

After basketball left the Coliseum, the court went into hibernation for several years, then was reinstated for use by the Nats when the American Bowling Congress tournament took over the War Memorial and left the cagers homeless for the NBA playoffs.

But surprisingly, after its near demise in the 1960's, if any activity has proven an off-season money maker for the fair, it has to be auto racing of the dirt track variety. Today Central New York and the fairgrounds constitute the world's center for this sport.

DIRT TRACKING

Even after the "big car" races were dropped following the 1962 fiasco, promoters were drawn to the track, one of the last remaining dirt miles in America.

The United States Auto Club brought in its sprint car division for twin 5-mile races on a bleak October day in 1969. Although few fans were there, the competition provided a showcase for such drivers as Johnny Rutherford and Gary Bettenhausen (son of the late Tony Bettenhausen, three-time Syracuse 100-mile winner) who won the National Sprint Car Championship that day.

A year later USAC's late model stock cars gave the track a try for a 100-mile championship. Again, attendance was low . . . the race a complete disaster. On the 62nd lap the run finished in a grinding crash in the backstretch. So many cars were wrecked that it ended the entire USAC stock car season and a number of drivers, including Indianapolis star Roger McCluskey wound up in the hospital.

They never returned to Syracuse.

It's doubtful if auto racing at the fair would ever again have been successful if it hadn't been for Glenn Donnelly, a fellow who had an idea dirt racing could be lifted from obscurity and relative notoriety into a well-paying business.

Donnelly, who had developed a dusty rinkydink track in a rural farm pasture to a major stop for weekly races, was ready to take on the state fair as a promotional project. Some people had discussed the possibility of building a 50,000-seat sports stadium for the fair track, with moveable bleachers enabling the track to have some functions. This stadium would also be used for Syracuse University football. However, when all the talking was over, what remained was a deteriorating plant with open bleachers painted a sad gray; a penchant for dust which caused generations of drivers to refer to it as "the pig pen," and a dramatic history.

At about this time another historic dirt track, the Langhorne Speedway near Philadelphia, closed down after serving as the home of the country's first important modified stock car races. These vehicles, having graduated from their jalopy days, had improved in performance to the point where they demanded a better showcase than the back pasture.

Donnelly seized the chance to move in as promoter of all fair auto racing events. Earlier drivers had risked their necks and equipment at the fairgrounds for purses of around $5,000, the winner often taking home $500. Donnelly was thinking in terms of winner's shares in the tens of thousands!

With the new promoter, everything changed. By 1972 he found an outside sponsor and was ready to go with the first "Schaefer 100" for modifieds. The response was unbelievable. Despite rain on the day trials were to be held, thus packing the two-day affair into a single afternoon, the open stands were filled. Racing cars arrived from throughout the northeast for the encounter. The winner was Buzzie Reutimann of Florida.[16]

By the following year the new grandstand was opened and visitors had the chance to enjoy races in comfort from a high enough elevation so action on the far side of the track could be seen. Previously events "over there" were pure guesswork for the press and many working officials as well as for spectators.

Donnelly also addressed himself to the considerable problem of dust. In earlier years the fair track was treated with water, soap, sawdust and oil in an effort to "keep it clean," but within a few laps the grooved dirt tires inevitably threw waves of sticky, choking dust and occasional large lumps of dirt into the stands. Even worse, the track would break up, making the surface extremely treacherous.

Heavy duty scrapers and graders were brought into play. The track would first be plowed up. Then tons of calcium chloride would be applied and blended with water. Next, hours were spent running the track with "junkers," cars equipped with heavy duty tires and loaded with weights.

The result was an almost dust-free surface that held together and gave the track the running characteristics of a paved circuit. It still, however, provided the tackiness essential to the broadslide turns that delight dirt track enthusiasts.[17]

Records for all classes of cars fell to the modifieds. Scores of talented drivers soon spread the word. A win at Syracuse could lead to NASCAR or USAC contracts.

In 1974 Donnelly felt comfortable enough to invite back the USAC "big car" 100-miler, changing the name to the Salt City 100. The 35th running of the race, first held in 1923, was set for July 4. A good field turned out, with Al Unser and Mario Andretti putting on a stirring charge until Andretti ran out of gas on the last lap.

For Gary Bettenhausen, the sentimental favorite, the race was anything but good. He was already in a Syracuse hospital following a freak accident in warm-ups. Going into the first turn, his car dug into a rut, then became airborne. It crashed into the roof of a trackside restaurant and lodged there, 10 feet above the outside rail, as firemen and ambulance attendants worked to free him. It was five months before Bettenhausen returned to racing.[18]

Failure of the big names to appear on a continuing basis spelled an end to the Salt City 100 in 1977 when Larry Dickson, a superb, though not too well-known driver, won.

Donnelly was not ready to give up on open-wheel competition. He added a unique "Supernational" event of 100 kilometers (63 laps) to the Schaefer Weekend, which was drawing 25,000 spectators by 1977. The "Supernationals" were designed for any type of open cockpit car including "big cars," sprinters and pavement-type supermodifieds, and unlimited horsepower class.

From a slow beginning this type of racing by 1980 was attracting the very best drivers through a series of qualifying races held in the east and midwest during the summer. The $13,332 first place money brought out 69 cars and drivers for time trials as they strove for one of the 40 starting positions.

They proceeded to break every existing record for flat, dirt mile tracks, with the magic 30-second mark smashed by Keith Kauffman in 29.628, an average speed of 121.506 mph in his Weikert Livestock Sprinter. The car's owner, Bob Weikert, celebrated by blowing $600 in a supper for 19 friends.*

The celebration came too soon. The car broke down in the race and Steve Kinser, driving a sprinter prepared by his father, managed to win the 100-kilometer event and the lion's share of the $60,000 purse.[19]

The Schaefer, expanded to 125 miles and renamed the "200" (for kilometers) and the Supernationals were joined by another racing attraction in 1982 when seven-ton supercharged truck tractors took to the oval for the first time in history. The brutes, belching streams of black diesel exhaust from highly-polished chrome stacks, were clocked at over 100 miles an hour in the straights. The 150-mile race was won by Charlie Baker of New Oxford, Pa. A world's record of 80.2 mph was set for the dirt mile by Baker in qualifying.

On the same day little JoJo Moratta, not yet 10 years old, ran a four-foot-long quarter midget around the mile for another world's record — 35.81 for the tiny car.

---

*   On October 5, 1984 Rocky Hodges of DesMoines, set a new world's record of 26.776 seconds, or 134.4487 miles an hour. During the race the following day he was out with a flat tire after 37 laps.

CHAPTER VIII

In Conclusion

The eighties were marked by change and litigation at the Fairgrounds.

In a major court action with Constitutional implications, Federal District Judge Howard G. Munson noted Hare Krishna (International Society for Krishna Consciousness) devotees must confine their fund raising activities at the fairgrounds to rented booths, rather than to wander about in their solicitation of funds, a Krishna ritual called "Sankirtan."

A similar finding was made in a case involving the Krishnas and a midwestern fair.

Krishna lawyers have contended this ruling violates their religious freedoms.

The state police were charged with roughly treating several persons apprehended at the 1983 fair, but a sergeant indicted by an Onondaga County grand jury on assault charges was exonerated.

The perennial political charges and countercharges of favoritism in the employment at the fairgrounds and in the letting of contracts came up again in 1984.

Meanwhile, close to the western edge of the race track a year-round village of harness horse trainers, drivers and their families had been unofficially established for several decades. The community of trailers, appropriately named "Horse" by *Herald-Journal* columnist Dick Case, came to an end in the summer of 1984, victim of antiquated utilities and changing times at the fair.

And the grounds have been improved by the addition of permanent lights around the one-mile track, which were first used on July 3, 1984. Most car drivers admitted their surprise at the good visibility, which also enabled spectators in the stands to identify cars in the farthest reaches of the oval.

There was only one problem. As the 30-car field came down the home stretch for the start of the first night's feature, one racer got sidewise. In seconds 15 others were involved in an expensive 120-mile-an-hour jumble. Although no one was injured, a dozen vehicles were missing at the restart.

Thus the fair continues to evolve in new and unique ways, yet remains very much of what it was first designed to be . . . officially a showcase for agriculture . . . and unofficially a time for fun between planting and harvesting.

Even its off-season functions represent events which are especially suited to its contours.

It would be hoped that eventual land development along the lakefront and true refurbishment of the existing buildings can enable the New York State Fair to assume its rightful place as the First Fair of the Nation.

A first step in this direction was made with the announcement on September 2, 1984 that a five-year plan is being considered by the state legislature and Governor Cuomo for a 50 to 70 million dollar capital improvement program. The entire infrastructure would be overhauled, adding to the ground's attraction as a year-round facility.

To this end, the governor's office and the State Commissioner of Agriculture, Joseph

Gerace, acknowledged consideration of an Agri-Mart plan for the grounds which would encompass a year-round multi-billion dollar agricultural market bringing in from two to three million visitors a year from around the world. As part of this, State Fair Director Tom Young proposes a summer-long recreational theme park as a tie-in. A plan on the drawing board brings back to life, after a quarter of a century, the idea of a marina for the adjacent lakeshore.

That the fair continues to be popular is evidenced by the record attendance of 767,585 people attending the 10-day 1983 event. This shattered the previous record of 714,025 set just one year earlier. The biggest day ever was achieved on Sunday, September 1, 1984, when 116,796 fairgoers went through the gates, eclipsing by 3,001, the old record set in 1970. It is believed that the saturation point for any single day has been reached, given the traffic, parking and on-grounds facilities now available.

Officials are contemplating a longer fair to counteract this problem, and to also achieve a better earnings picture.

But Young put things in prespective when he cautioned that mere numbers of people through a turnstile are not the sole measures.

"The barometer," he concluded, "is how people feel when they leave the gate. I think this year's fair left people feeling uplifted."

If so, then they'll be back.

Bibliography

Armstrong, George, *Forestry College, Essays on the Growth and Development of the New York State's College of Forestry*, Syracuse, 1961, 360 pp.

Baker, Anne Kathleen, *A History of Old Syracuse, 1654-1899*, Manlius Publishing Company, Fayetteville, N.Y., 1937.

Beauchamp, Rev. William M., *Onondaga Historical Association Annual Volume, 1914*.

_____, *Past and Present of Syracuse and Onondaga County*, S.J. Clarke Publishing Co., Inc., 1908.

Beveridge, William, *Influenza: The Last Great Plague*, Prodist, New York, N.Y. 1977, 124 pp.

Brewster, Arthur Judson, *Memories of Clinton Square and Other Tales of Syracuse*, Syracuse University Press, Syracuse, 1951.

Bruce, Dwight H., *Memorial History of Syracuse*, H.P. Smith & Co., 1981, 718 pp.

Catalogue of Entries for the Fifty First Annual Cattle Show and Fair of the New York State Agricultural Society at Syracuse, 1891, Van Benthuysen Printing House, Albany, 400 pp.

Chase, Franklin H., *Syracuse and Its Environs* (3 vol.), Lewis Historical Publishing Co., Chicago, 1924.

Chemung County Historical Journal, Chemung County Historical Society, Elmira, N.Y., 1962.

Clayton, W.H., *History of Onondaga County*, D. Mason & Co., Syracuse, 1878, 430 pp.

Collier, Richard, *Plague of the Spanish Lady*, The Antheneum, 1974.

Collier's *Encyclopedia*, 1980.

Cooley, Miriam, *The First 100 Years — A History of the Syracuse Chapter of the American Red Cross*, Manlius Publishing Co., Fayetteville, 1981, 167 pp.

Crosby, Alfred W., Jr., *Epidemic and Peace 1918*, Greenwood Press, Westport, Conn., London, England, 1976, 337 pp.

DuVall, John, *Years of Our Youth*, Unpublished manuscript, Syracuse, 1961, 401 pp.

Encyclopedia Americana, 1980.

Encyclopedia International, 1979.

Estey, Emily, *Papa Was Postive*, Heritage Press, Sherburne, N.Y. 1966.

Fairs, U.S.A. — New York State, Image Digest, Inc., Ithaca, 1970.

Fellows, Byron F., Jr., and Roseboom, William F., *The Road to Yesterday*, Manlius Publishing Co., Fayetteville, N.Y., 1948, 103 pp.

Galpin, W. Freeman, *Central New York — An Inland Empire* (3 vol.) Lewis Publishing Co., Inc., Chicago, 1941.

Gater Racing News Yearbook — 1973, 1975, 1979. Sports Trek, Inc., Syracuse, 1973, '75, '79.

Georgano, G.N., *Encyclopedia of Motor Sports*, Viking Press, New York City, 1971, 656 pp.

Greater Syracuse Chamber of Commerce Membership Directory, 1980-81, Syracuse.

Hand, H.C., *Syracuse From a Forest to a City*, Master & Stone, Syracuse, 1889.

Hedrick, Ulysses Prentiss, *History of Agriculture in the State of New York, A*, Hill & Wang, (American Century Series), 1966, 466 pp.

Hoehling, Adolph, *The Great Epidemic*, Little, Brown, Boston, 1961.

Howe, John B., *The New York State Fair, Its Genesis and Its History*, Hall & McChesney, Syracuse, 1917, 76 pp.

Illustrated Annual and Program of the New York State Fair, The Catholic Sun, Syracuse, 1892.

Invitation from the State of New York to Hold Seventh World's Poultry Congress at the New York State Agricultural and Industrial Exposition Grounds, Syracuse, 1937.

Keegan, J.J., *The Prevailing Pandemic of Influenza, Journal of the American Medical Association*, Vol. 71 (Sept. 28. 1918) p. 1051.

Macdonald, Rod, *Syracuse Basketball, 1900-1975*, Syracuse University Press, Syracuse, 1975, 140 pp.

McKinley, J.W., *Assassinations in America*, Harper & Row, New York 1975-77, 243 pp.

Melone, Harry R., *History of Central New York* (3 vol.) Historical Publishing Co., Indianapolis, Ind., 1932.

Morgan, Howard, *William McKinley and His America*, Syracuse University Press, Syracuse, 1963.

Mystique Krewe Programs, 1905-1910, Syracuse.

New and Greater New York State Exposition: A Presentation to the Honorable Thomas E. Dewey, Governor of New York State From The People of Syracuse and the Syracuse Chamber of Commerce. Syracuse, 1946, 32 pp.

New York — A Guide to the Empire State — American Guide Series, Eighth Printing; Oxford University Press, New York, 1962, 752 pp.

New York State Agricultural Society, Proceedings of the Annual Meeting, 1841-1900, Albany.

New York State Fair, Syracuse, 1913.

O'Brian, William, *Forty Years on the Force*, Syracuse Herald, Syracuse, 1926.

Palmer, Richard R., *The Old Line Mail*, North Country Books, Lakemont, New York, 1977, 172 pp.

Premium Book, 1921 New York State Fair, 1921, 188 pp.

Premium Book, 1922 New York State Fair, 1922, 184 pp.

Profile of Central New York, Syracuse Governmental Research Bureau and Metropolitan Development Association, Syracuse, 1975, 139 pp.

Rappoport, Ken, *The Syracuse Football Story*, The Strode Publishers, Huntsville, Ala., 1975, 352 pp.

Roseboom, William F., and Schramm, Henry W., *They Built a City*, Manlius Publishing Co., Fayetteville, N.Y., 1976, 178 pp.

Salina, A Business History, Syracuse Business.

Schramm, Henry W., *The City Built on a Pillar of Salt*, Unpublished manuscript, 1955.

_____, and Roseboom, W.F., *Syracuse — From Salt to Satellite*, Windsor Publications, Woodland Hills, Calif., 1979, 244 pp.

Shaw, Wilbur, *Gentlemen, Start Your Engines*, Coward-McCann, Inc., New York, 1955, 320 pp.

State Fair Reports, New York State Department of Agriculture, 1853, '91, '95, '98, 1910, '13, '17, '20, '21 and '22.

Turkin, Hy and Thompson, S.C., *The Official Encyclopedia of Baseball, Second Revised Edition*, A.S. Barnes and Company, New York, 1959, 593 pp.

Thurlow, Walter R., *Illustrated Annual & Program of the New York State Fair*, Syracuse, 1895, 104 pp.

United States Auto Club Yearbook, Indianapolis, Ind., 1959.

Watson, Elkanah, *New York State Agriculture, the Rise, Progress and Existing State of Modern Agricultural Societies of the Berkshire System*, D. Steele, Albany, 1820, 208 pp.

Welch, Walter L., *Past and Future of the New York State Agricultural and Industrial Exposition*, Syracuse, 1946, 64 pp.

World Book Encyclopedia, 1980.

NEWSPAPERS AND PERIODICALS

Area Auto Racing News
Camillius Enterprise
D.I.R.T. Auto Race Programs, 1973-1981.
Gater Racing Photo News
Herald, The
Herald-Journal, The
Herald-American, The
Jordan Leader
Journal, The

Marcellus Observer
National Speed Sport Weekly
New York State Fair Programs, 1900-1983.
New York Tribune
Parade Magazine
Post-Standard, The
Syracuse *Daily Standard*
Tully Times

Notes

CHAPTER I   ON THE ROAD

1. Schramm, H.W. and Roseboom, W., *Syracuse — From Salt to Satellite*, pp. 6, 7.
2. *Sunday Herald*, October 9, 1898.
3. Watson, Elkanah, *New York State Agricultural History*, p. 118.
4. Ibid., pp. 118-137.
5. Ibid., pp. 148-155.
6. Welch, W.L., *New York State Agricultural and Industrial Exposition*, p. 10.
7. Howe, J.R., *New York State Fair, Its Genesis and History*, pp. 7, 8.
8. Ibid., pp. 9-13.
9. New York State Agricultural Society, *Proceedings of the Annual Meetings, 1841*, p. 15.
10. Chase, F.H., *Syracuse and Its Environs*, p. 503.
11. *State Society Proceedings, 1841*, p. 30.
12. Ibid.
13. Howe, pp. 9-13.
14. Palmer, R., *The Old Line Mail*, pp. 124-132, 165.
15. *State Society Proceedings, 1841*, p. 43.
16. Ibid.
17. *Sunday Herald*, October 9, 1898.
18. *Herald*, August 5, 1928.
19. *Report of the New York State Fair, 1910, Bulletin 25*, pp. 502-507.
20. Ibid.
21. Howe, pp. 13-18.
22. *Sunday Herald*, October 9, 1908.
23. *The Post-Standard* (date unknown).
24. *State Society Proceedings, 1854*, pp. 12, 13.
25. Ibid., pp. 12-17.
26. *State Fair Report, 1910*, pp. 502-507.
27. *Chemung Historical Journal*, December, 1962, pp. 1039-1044.
28. Ibid., pp. 1043, 1048, 1049.
29. *Herald*, September 8, 1907.
30. *State Fair Report*, 1910, pp. 502-507.
31. *Daily Standard*, October 7-11, 1858.
32. *Daily Journal* (date unknown)
33. Howe, pp. 18, 19.

CHAPTER II   THE FAIR COMES HOME

1. Chase, F.H., *Syracuse and Environs*, p. 524.
2. *The Post-Standard*, September 13, 1903.
3. Howe, J.B., *New York State Fair, Its Genesis and History*, p. 19.
4. Chase, p. 524.
5. *The Standard*, September 12, 1890.
6. Ibid., September 13, 1890.
7. *The Post Standard*, July 20, 1948.
8. *1892 Illustrated Annual and Program*.
9. *The Post-Standard*, September 21, 1893.

10. *1893 Illustrated Annual and Program*.
11. *The Post-Standard*, September 22, 1893.
12. Rappoport, Ken, *The Syracuse Football Story*, pp. 20, 21, 333.
13. Pierce, Enid C., State Fair of 90's, *Herald-Journal*, September 9, 1945.
14. *Tully Times*, September 9, 1954.
15. *1898 Illustrated Annual and Program*.
16. Welch, Walter L., *New York State Agricultural and Industrial Exposition*, pp. 11, 12, 22, 23.
17. Ibid., p. 23.
18. Chase, p. 525.
19. *New York State Agricultural Society, Proceedings of the Annual Meeting*, 1900, p. 2.
20. Ibid.
21. Welch, p. 13.
22. *State Society Proceedings*, 1900, p. 3.
23. Roseboom, William H., and Schramm, H.W., *They Built a City*, p. 162.
24. Leech, M.K., *In the Days of McKinley*, pp. 594-603.
25. McKinley, J., *Assassination In America*, p. 55.
26. *The Post-Standard*, September 11-13, 1901.
27. *The Post-Standard*, September 14, 1901.
28. Ibid.
29. *The Post-Standard*, September 15, 1901.
30. *The Journal*, September 7, 8, 1903.
31. *The Post-Standard*, September 8, 1903.
32. Ibid.
33. Roseboom and Schramm, p. 164.
34. *The Post-Standard*, September 14, 1903.
35. O'Brian, W., *Forty Years on the Force*, pp. 96-98.
36. Chase, p. 526.
37. *Program, Mystique Krewe*, 1904.
38. Beauchamp, William M., *Past and Present of Onondaga County*, pp. 470, 471.
39. *Marcellus Observer*, September 1, 1905.
40. *The Post-Standard*, September 19, 1905.
41. *Camillus Enterprise*, September 22, 1905.
42. Welch, pp. 13, 15.
43. *The Journal*, September 12-15, 1910.
44. DuVall, John C., *Years of Our Youth*, pp. 261-272.

CHAPTER III   THE DAYS OF THE PALE HORSE

1. *The Post-Standard*, September 18, 1911.
2. *The Post-Standard*, September 17, 1931.
3. *The Post-Standard*, September 12, 1934.
4. *The Post-Standard*, September 10, 1912.
5. *The Journal*, September 11, 1912.
6. *The Post-Standard*, September 10, 1912.
7. Howe, J.B., *New York State Fair, Its Genesis and History*, pp. 9-12.
8. Fairman, Roy E., As Time Goes By, *Herald-Journal*, 1957.
9. *The Post-Standard*, August 26 - September 18, 1916.
10. *The Journal*, August 26 - September 18, 1916.
11. *The Post-Standard*, September 12, 1916.
12. Chase, F.H., *Syracuse and Environs*, p. 546.
13. Cooley, M., *The First 100 Years, A History of the Syracuse Chapter, American Red Cross*, pp. 13, 15, 19.
14. *Marcellus Observer*, August 17, 1917.
15. Howe, p. 27.

16. *The Post-Standard*, September 14, 1917.
17. Chase, pp. 558, 559.
18. *The Journal*, May 26, 1918.
19. *The Post-Standard*, September 5-14, 1918.
20. Crosby, A.W., Jr., *Epidemic and Peace*, pp. 39, 40.
21. Collier, Richard, *Plague of the Spanish Lady*.
22. *The Post-Standard*, September 21, 1918.
23. Cooley, pp. 26-30.
24. *The Post-Standard*, September 14 - October 25, 1918.
25. *The Herald*, September 14 - October 14, 1918.
26. Crosby, pp. 319, 320.

CHAPTER IV   FLAMBOYANT YEARS

1. Estey, Emily, *Poppa Was Positive*, p. 128.
2. *The Post-Standard*, September 7, 1919.
3. *The Post-Standard*, September 19, 1920.
4. *The Journal*, September 19, 1920.
5. *The Post-Standard*, February 15, 1923.
6. *The Journal*, October 5, 1923.
7. *The Post-Standard*, October 6, 1923.
8. *The Post-Standard*, September 2, 1924.
9. *The Post-Standard*, September 11, 1924.
10. *The Journal*, September 16, 1924.
11. *The Post-Standard*, September 16, 1924.
12. *The Journal*, August 30, 1926.
13. *The Post-Standard*, August 31, 1926.
14. *The Post-Standard*, August 28, 1928.
15. *The Journal*, August 25, - September 8, 1929.
16. *The Post-Standard*, August 25 - September 8, 1929.
17. *The Post-Standard*, September 5, 1929.
18. *The Post-Standard*, September 18, 1929.
19. Roseboom & Schramm, *They Built a City*, p. 167.
20. *The Post-Standard*, August 24, 1966.
21. Welch, W.L., *New York State Agricultural and Industrial Exposition*, pp. 17-21.
22. Ibid.
23. *Marcellus Observer*, September 5, 1934.
24. *The Post-Standard*, January 24, 1937.
25. *State Fair program flyers*, 1938, 1939.
26. Armstrong, George, *Forestry College Essays*, pp. 264, 265.
27. *The Journal*, September 12, 1931.
28. *Herald-American*, September 10, 1934.
29. *The Post-Standard*, September 11, 1934.
30. *The Post-Standard*, September 9, 1934.

CHAPTER V   A TIME TO REBUILD

1. Cooley, M., *Our First Hundred Years, Syracuse Chapter, American Red Cross*, p. 55.
2. *The Herald-Journal*, July 29, 1946.
3. *New and Greater New York State Exposition, A*, Presentation to Governor Thomas E. Dewey, 1946.
4. Welch, W.L., Letter to the Editor, *The Post-Standard*.
5. *The Herald-Journal*, January 17, 1948.
6. *The Post-Standard*, September 17, 1948.

7. *The Post-Standard*, December 31, 1948.
8. *The Herald-Journal*, September 5, 1949.
9. *The Post-Standard*, September 6, 1949.
10. *The Post-Standard*, May 28, 1950.
11. *The Post-Standard*, September 11, 1954.
12. *The Post-Standard*, September 7, 1936.

CHAPTER VI   THE MODERN ERA

1. *The Herald-Journal*, September 4, 1959.
2. *The Post-Standard*, September 5, 1959.
3. *The Herald-Journal*, August, 1961.
4. *The Post-Standard*, June 24, 1962.
5. *The Herald-Journal*, August, 1962.
6. *The Herald-Journal*, August 31, 1963.
7. *The Herald-Journal*, January 16, 1967.
8. *The Post-Standard*, September 2, 1968.
9. *The Herald-Journal*, August 26 1969.
10. *The Herald-Journal*, August 26, 1975.
11. *The Herald-Journal*, September 2, 3, 1975.
12. *The Post-Standard*, September 3, 1975.
13. *The Post-Standard*, September 27, 1975.
14. *The Herald-Journal*, August 26, 1978.
15. *The Post-Standard*, July 12, 1976.
16. *The Post-Standard*, August 30, 1980.
17. *Empire Magazine, The Herald-American*, August 20, 1978.
18. *The Herald-American*, August 24, 1980.
19. *The Post-Standard*, August 25, 1980.
20. *The Post-Standard*, August 28, 1980.
21. *The Post-Standard*, August 22, 1980.
22. *The Herald-Journal*, August 23, 1980.

CHAPTER VII   OFF-SEASON

1. *The Herald-American*, July 13, 1980.
2. *The Herald-Journal*, July 3, 1980.
3. Donald J. Wickham letter to Governor Nelson Rockefeller.
4. Governor Nelson Rockefeller letter to Donald J. Wickham, October 21, 1970.
5. *Hambletonian Presentation Brochure*, 1973.
6. *The Herald-Journal*, October 6, 1979.
7. *The Herald-American*, August 17, 1980.
8. *Book of the Century (Centennial Edition), Syracuse Journal*, 1939.
9. *Greater Syracuse Chamber of Commerce Directory*, 1980. pp. 58, 59.
10. Roseboom, W. and Schramm, H.W., *They Built a City*, pp. 154, 155.
11. Macdonald, Rod, *Syracuse Basketball*, pp. 46-55.
12. *Chamber Directory*, 1980, pp. 54, 55.
13. *The Herald-Journal*, November, 1949.
14. *The Post-Standard*, November, 1949.
15. *Syracuse Nationals Game Programs*.
16. *Schaefer 100 Yearbook, 1973*.
17. *Gater Racing News*, September 29, 1978.
18. *The Herald-Journal*, July 5, 1974.
19. *Gater Racing News*, October 17, 1980.

www.ingramcontent.com/pod-product-compliance
Lightning Source LLC
Chambersburg PA
CBHW080346300426
44110CB00019B/2524